'If I were "king for a day", Avery would [   ] benign dictator of conservation in the U. [   ] this, think about this, and act upon this.'

Chris Packham, broadcaster and author of *Back to Nature*

'A timely, brutally honest, yet inspiring account on what has gone wrong with wildlife conservation, and how we can put it right.'

Stephen Moss, naturalist and author

'Mark Avery has been a guiding light in conservation all my life; a constant north star. This important book bears witness to what we've lost, what we've done about it, what works and what we must do next. It is both a reckoning and a resounding call to real action, at the most crucial time of our lives – of all our wild lives. Here is hope, predicated on action. There is work to do; and we'd better get on with it.'

Nicola Chester, RSPB columnist and author
of *On Gallows Down: Place, Protest and Belonging*

'*Reflections* is a work of distilled campaigning wisdom, told with the irrepressible optimism of a passionate advocate for nature who's spent decades working tirelessly for wildlife. With wit, verve and clarity of prose, Mark Avery lays out a strikingly radical set of proposals for how to turn around the decline of wildlife in these isles.'

Guy Shrubsole, environmental campaigner and author
of *The Lost Rainforests of Britain* and *Who Owns England?*

'Mark Avery has written a love letter to Nature. Yes it is well written and academically sound and all that you'd expect from a person of his track record, but the real pleasure of the book is that under all that patina of propriety and science you feel a Mr Darcy launching himself into the lake because nothing is more important to him than capturing our hearts with his passion. A real triumph.'

Sir Tim Smit, Co-founder and Vice Chairman
of the Eden Project

'Dr Avery must be congratulated on this important book. He hits the nail on the head. I found myself nodding my head vigorously while reading it. The time for action is now.'

Baron Randall of Uxbridge, RSPB Council member and peer

'Mark Avery is uniquely qualified to write this immensely stimulating and thought-provoking book. Reflecting on his lifetime in conservation he discusses the successes and failures of the past, and draws important lessons for more effective conservation in the future.'

Professor Ian Newton FRS, ornithologist and conservationist

'A clarion call for more nature in Britain and how we can get it. Wise, knowledgeable, provocative and good humoured – Mark Avery is a national treasure.'

Patrick Barkham, author of *Wild Green Wonders*
and co-author of *Wild Isles*

'A brilliant, thorough book full of insightful observation. A must read for those who care about natures future and wish to understand the character of our contorted relationship with it.'

Derek Gow, author of *Bringing Back the Beaver*

'Deeply felt and clear eyed, this book admirably achieves its aim of being "realistically hopeful" about a wildlife renaissance and what it will take for us to get there. You don't have to agree with all its conclusions. But the questions it intelligently explores, based on a lifetime of experience in conservation, of "what sort of world do we want to live in?" and "what should I do about it, then?" are the essential ones of our times. Read it and be both enlightened and challenged.'

Beccy Speight, CEO, RSPB

REFLECTIONS

REFLECTIONS

What wildlife needs and how to provide it

MARK AVERY

PELAGIC PUBLISHING

Published in 2023 by
Pelagic Publishing
20–22 Wenlock Road
London N1 7GU, UK

www.pelagicpublishing.com

Reflections: What wildlife needs and how to provide it

A CIP record for this book is available from the British Library

ISBN 978-1-78427-390-3 Paperback
ISBN 978-1-78427-391-0 ePub
ISBN 978-1-78427-392-7 PDF
ISBN 978-1-78427-447-4 Audio

https://doi.org/10.53061/UMDF3032

Typeset in Minion Pro by S4Carlisle Publishing Services, Chennai, India

To Megan and Renée, great-grandmothers born in 1926, and James, their great-grandson, born in 2021

Contents

Preface

This book is about wildlife and wildlife conservation in the UK. It is grounded in the early 2020s but looks back over recent decades and forward to decades to come. Since I am in my sixties, I have more experience and action behind me than ahead of me, but the aim of this book is to give a nudge to a better future for UK wildlife by giving the reader an insider's overview of the challenges facing those who wish to help wildlife to thrive. It is, in essence, a hopeful book, but a realistically hopeful book, which gives pointers to what needs to be done and some tips on the part readers themselves might play in delivering a better future for UK wildlife. There are no quick fixes; there's a lot of hard graft ahead.

There is no way that a book of 85,000 words could be comprehensive in its treatment of the delights we find in wildlife, the problems it faces, and how we could give it more of what it needs. Well, if there is, such a book is well beyond my capabilities. Instead, I have tried to provide enough case studies and explanation in each chapter to carry the reader forward into the next wanting to know how the tale unfolds.

The book has six chapters, each including my personal reflections at its end. There is a narrative trajectory which starts with local observations of wildlife around my home as an introduction to some species and to some issues. My day-to-day relationship with wildlife in the house, in the garden and close by leads me to ponder the overall state of UK wildlife – that's the subject for Chapter 2. There is no escaping the conclusion that our wildlife has been in long-term decline for many years, in fact for centuries. Maybe this is an inevitable consequence of human so-called progress, with our material wealth increasing as wildlife richness declines. If so, we ought to take a close look at the aims and objectives of wildlife conservation, an enterprise in which I have spent most of my life, because it doesn't seem to be living up to its name very well. That is the subject of Chapter 3. How does looking after wildlife fit in to the way that we see the future? Lest this seems like a very gloomy road to travel, I believe that wildlife in the UK is far from doomed, and there are plenty of reasons to think that we can do much better over the next few years. Chapter 4 highlights wildlife success stories from conservation policy and practice which demonstrate the scope for a wildlife renaissance if only we get things right. Chapter 5 looks at ways in which we aren't getting things right at the moment, and in Chapter 6 I set out my thoughts on what we should do to make things better. Wildlife decline is a problem caused by our society, and the achievement of a wildlife recovery will have to be a shared achievement.

Mark Avery

Some explanations

Wildlife or nature?

The distinction between nature and wildlife is a fine one, but my view is that nature is a broader concept that includes wildlife. Nature, for me, and there is support for this view in many dictionaries, includes landscape, geology and geomorphology, and that's not what this book is about. Wildlife itself isn't a completely clear and unfuzzy concept but, in this book, it refers to non-domesticated animals and plants, fauna and flora. And I haven't used the term biodiversity because it's a ghastly word.

If you were to search for the word 'wildlife' in this book you would find it over 1,000 times and discover that I refer to wildlife conservation and wildlife reserves where others might have used nature conservation and nature reserves. Search for 'nature' and you'll find it fewer than 100 times, and those are mostly in names such as the Nature Conservancy Council and in phrases such as 'in the nature of things'. 'Biodiversity' occurs only where others have used it and I am quoting them. There are a lot of instances of the word 'wildlife' in these pages, but it's a pretty good word, I think. In meaning 'living things that are in the wild' it does a very good job in eight letters.

Species names

I have only very rarely used scientific names in this book. I just don't think that telling you the scientific name for the Nightingale is going to make this a better book – so I haven't. But it is *Luscinia megarhynchos*.

I have not stuck to any list of authorised species names. I've used whichever vernacular name I usually use in conversation. I don't think this will lead to any confusion, but if it does – sorry! This means that I write about Nightingales but don't refer to them as either Common Nightingales (they are getting rarer, after all) or Rufous Nightingales (it's usually too dark to tell how rufous they are).

When it comes to writing the English-language names for species, there are two options: capitalise species names or not. I have a marginal preference, and an ingrained habit as a result, of starting species names with capitals, as in Common Gull and Shy Albatross, for the reason that not flagging them up as species in this simple way can lead to people thinking that one is writing about all those gulls that are common and all those timid and bashful albatrosses (those are real examples). I fully accept that this is a personal choice, and one which different individuals and organisations differ on. I worked for the RSPB for 25 years and never capitalised a species name in all that time. I also accept that my preferred method leads to some ugly text on occasions where a list of creatures contains some species (capitalised) and some groups of species (not capitalised) – for an example, see a sentence in the second paragraph of the text about driven grouse shooting in Chapter 4. This issue may seem to be of Lilliputian egg-opening proportions but you'll be able to see from the book whether the author was overruled by the editor/publisher combination.

Devolution

I am a fan of devolution, which has resulted in each of the four UK nations having power over many matters within its geographical borders. These include areas central to this book such as wildlife con-servation, farming, forestry, fisheries and planning controls. And that means that since 1997 the details of laws and policies and practice have tetraverged (or maybe quadriverged) so that things are a bit, or a lot, different in different UK nations. This is a pain for someone writing a book, because one has either to explain all four situations or say that things are a bit different everywhere. I have tended to describe the situation in England, not just because I am English and live in England and understand the situation better for England, but also because 80% of the population of the UK live in England and it forms well over half of the UK land area. I have resorted to 'things are a bit different elsewhere' quite a lot, but I have also chosen, deliber-ately, to highlight certain areas where things are very different in the UK's four nations, and to choose examples from outside of England whenever I was fairly confident that I knew what I was talking about – more commonly in biology than in politics.

There are five, not four, national governments in the UK: the Welsh, Northern Irish and Scottish governments and the UK government in London – because some matters such as taxation (most of it), defence, foreign affairs, immigration and some energy policy are not devolved – and what is essentially an English government which deals with the devolved issues in England, which is based in London too. Nobody really says 'English government', probably because the cabinet ministers around the table in Downing Street have a mixture of UK and English roles, so the same people are a UK government and an English one. I have tried to make the situation clear by the use of the word governments (plural) when I'm talking about all of them, and using the titles Northern Irish, Welsh and Scottish when I am talking about those in particular. When dealing with English matters I have tended to name the government department with responsibility, most often the Department for Environment, Food and Rural Affairs (Defra), which has some UK responsibilities, particularly in international matters, and English responsibilities too.

Notes and references

The notes, references and suggestions for further reading at the end of this book have been given quite a lot of thought (so I do hope you glance at them). They provide links to sources of information used in the text here but also some places to go to learn more – and they occasionally contain comments on who deserves credit for successes which were achieved or blame for those which were not.

CHAPTER 1

Glimpses of wildlife

Our collective response to wildlife is the sum of around 68 million people's relationships with wildlife. Some of that is played out in our own personal reactions to wildlife around us, whether we shoot it, spray it, put food out for it or ignore it completely. But, as we shall see, many of the critically important parts of our collective relationship with wildlife are at arm's length, through the actions of government, businesses, local councils and conservation organisations. Those relationships are ours too because they depend on how we allow our taxes to be spent, how we spend our money on goods and food, and the extent to which we support conservation charities to do their work. The UK's relationship with wildlife is not just about how many of us want to cuddle a Badger and how many of us want to cull one, but about whether the structure of agriculture as a whole is favourable to Badgers and a host of other species – and the same is true of forestry, fisheries, housing policy, transport policy and a plethora of other societal decisions.

I've been working in wildlife conservation for 35 years but have been passionate about wildlife for 55 years or more. That passion for wildlife remains undimmed over the decades and has suffused my working life. It is still with me, and every day is an opportunity to notice birds, plants and insects. Those continuing encounters with wildlife are still the emotional basis for much of my conservation work, and the same is true of many other wildlife conservationists I know. And so that's where this book starts, with wildlife on my doorstep and in my neighbourhood. By telling you about my own experiences of wildlife I will begin gently to explore some wider issues and themes. But also, it seems wrong to start a discussion of wildlife conservation without a quick look – and this is my personal look – at wildlife close to home.

Herb-Robert

At 6 a.m. I'm usually making the first pot of tea of the day after spending an hour or so sitting at a computer. On Mondays, Wednesdays and Fridays I'll fetch the milk bottles, which arrived hours earlier, from the front door while the kettle boils.

As I open the front door, I get my first feel of the outside world. If cold or wind or rain doesn't immediately drive me back indoors then I usually pause, and look at the sky, listen for birds and take another four paces to the curtilage of my property and look right to the church spire to the north, forwards down Clare Street and left along the road. In late spring I look left and down and see a Herb-Robert plant growing from the crack where my short wall meets the pavement.

Herb-Robert grows there most years, not quite every year, but most years. Every year there are Forget-me-nots, and this year there is Groundsel too, and a non-native but naturalised Red Valerian. The Herb-Robert is my favourite.

Herb-Robert is an annual or biennial native flower with five red petals and jagged leaves. It's a geranium or cranesbill. The Woodland Trust describes it as a 'low-growing plant with pretty red flowers'. Spot on! They tell me it's often found 'growing in the shade of woodland edges, next to walls and in other darker spots', so they seem to know their stuff. But the Royal Horticultural Society describes it as a weed and goes on to denigrate it some more by stating that 'it can be a useful ground cover but its tendency to self-seed can make it a nuisance as a garden weed', and that the time to 'treat' it is spring–autumn (which is pretty much all the time it shows above ground).

We have here a real dichotomy in how we view wildlife. Some regard it as a boon, others as a bane. Some look to wildlife as a wonder, others as a nuisance which should be 'treated'. That the allegedly plant-loving RHS's website is littered with name-calling of native plants, and advice on when and how to 'treat' them, is one of the more striking examples I have come across of different perspectives on the native wildlife around us. To be fair, we all have something of this dichotomy in us, we're strung out on a continuum, but the RHS, I now notice, is much further towards one end than I would have expected.

Later in the day I visit the post office, and use that as a chance to search many neighbouring streets for Herb-Robert. I walk with my head down and that is unusual for me, a birdwatcher, but I am searching the pavements and gutters for Herb-Robert. These are the quiet streets of a former shoe-town, Raunds, where there were, from Victorian times until the 1980s, many small workshops and factories producing footwear. Now the shoe works have gone, the shoes are made in East Asia and Raunds functions as a dormitory town for workers in Northampton, Peterborough and Bedford. The post office is in Brook Street, where there is indeed a brook, although most of its course is invisible and underground. It is the main street, where as well as the post office there is a Methodist chapel, a charity shop, a Co-op supermarket, a newsagent, a Chinese takeaway and more, but no Herb-Robert that I can see.

I turn, just before what is alleged to be the shortest pedestrian crossing in the country, and by the best fish and chip shop in town, and climb Hill Street before turning back into my street, and look at the other house fronts where they join the pavement. Some are plant-free, but one has a good line of Thale Cress and several have Shepherd's Purse. My neighbour has mostly Forget-me-nots and a Red Dead-nettle, but no Herb-Robert. There really are no Herb-Roberts along the other frontages. Mine is the Herb-Robert house in Lawson Street – a fact about which I am ridiculously pleased.

The quiet wood

I had meant to visit a few days earlier, but the weather forecast put me off. As I park by the wood, the rain has stopped and there are no other cars. There are no other cars because this wood, locally renowned for its Nightingales, has not heard the song of the Nightingale in two of the last three years, and nor have they been heard, so far, this year. As I walk into the wood, my bat detector tells me that the bat feeding above my head is a Soprano Pipistrelle, and my ears tell me that there are Song Thrushes, Blackbirds and Robins singing.

This is a wildlife reserve of the local Wildlife Trust and it looks to me every bit as suitable for Nightingales now as it did in the 1980s, but admittedly I am no Nightingale. I have come here with my parents (and Dad died over 25 years ago), with my young children, and with those same offspring now that they are adults. It has been a part of my life and a part of our family life for all that time. It has been a milestone of the year – hearing the Nightingales.

I wonder how long the Nightingales have sung here. Peter Scott was at school in the 1920s in nearby Oundle, an easy cycle ride away, and a line drawing from his early years was of a Nightingale singing. I wonder whether he had this wood in mind when he drew it, and whether he made the journey here, and stood near to where I am standing now and listened to Nightingales throwing their songs into the wood as day turned into night.

This evening, the Song Thrushes are on good form – they almost always are. I've sometimes noticed visitors leaving this wood after listening to a wonderful Song Thrush thinking they have heard a wonderful Nightingale and have considered whether to advise them to hold on a bit longer but decided, as much through shyness as kindness, to let them pass in blissful ignorance. Normally, the two species overlap in time, almost as though the Song Thrushes are handing over to the Nightingales for the nightshift.

In earlier years, it was a common experience to wait and wait for the Nightingale only for one to burst into song, loud song, within a metre of where we were standing. Suddenly a bramble thicket, which had been brown and green when we arrived, but was now black in the fading light, would reveal that it held one of our finest songsters. But not tonight. As the Song Thrushes subside into quietness nothing comes to replace them except a Woodcock flying over calling.

I feel this loss personally – these have been our Nightingales. Yes, we have shared them with many others, and gladly, but they have been ours. We have built them into our annual calendar and they have been a shared experience in our family. It looks as if I won't be bringing my grandson, James, here to hear them as his mother and uncle were brought here. And since he currently lives further north, in a county with Ring Ouzels but no Nightingales, I'm not sure who is going to provide James with his dose of Nightingale culture. There will be other wildlife experiences for him, but this thread, a family thread, is broken. And that makes me sad. It makes me much sadder than I would have expected.

Eventually, long after I would normally have heard the song if the birds were here, I give up and leave. Everything has seemed just the same as usual except for the absence of the star bird. The dusk chorus was just as usual, and as it subsided it became easier to hear the cries of lambs in the fields outside the wood and the distant calls of Tawny Owl, Pheasant and Muntjac Deer inside it. But the star bird was absent, and this has turned from being a possible one-year blip to a new reality. We can cross this wood off the list of Northamptonshire woods with Nightingales. And the same is happening in a growing number of other woodlands across England. Nightingales are slipping away, maybe not terminally, but seriously.

Red Kites

When I first moved to Raunds, about 35 years ago, the sight of a Red Kite would be hot news. Now, they are seen every day that one looks. It is a comeback, a recolonisation rather than a colonisation. In the avifauna of Northamptonshire, published in 1895 by the fourth Baron Lilford, the Red Kite was described as having been lost from the county due to habitat loss and the (then legal) activities of gamekeepers. Lilford recalled, as a young boy, so this was probably in the late 1830s or early 1840s, standing on the lawn at Lilford Hall and seeing three Red Kites overhead – but that was a rare sighting by then. Now, if you drive east from here towards Oundle, you will see Red Kites over the fields around Lilford as a commonplace.

As a teenage birder living near Bristol in the 1970s, I would not have imagined sitting in a Northamptonshire garden and getting great views of Red Kites flying overhead, and hearing them calling, every day. Back then, there were around 50 pairs of kites nesting in the UK and all were in mid-Wales in the upland sheep country of counties that are now Dyfed and Powys. We holidayed in mid-Wales more than most, since my mother is Welsh, and so eventually I saw my first Red Kite, floating in the far distance over a Welsh hillside, and I watched it through my first pair of binoculars until my eyes watered, so that I could be absolutely sure that what I thought was a forked tail really was properly notched. I couldn't see that the tail was orange, nor that the underwing had white and black patches, but I could see that the bird flew differently from a Buzzard, with its longer wings drooped rather than upturned and with an air of ease and mastery of the skies which made a Buzzard look a bit stolid. Now I can often see orange tails, without binoculars, by glancing up when a Red Kite flies over my garden at rooftop height.

Red Kites were brought back to English and Scottish skies, and subsequently to Northern Ireland too, through a chain of reintroduction sites from the north of Scotland to the Chilterns of England, one of which was in Northamptonshire. They prospered, and that's why I see them daily.

My children, now adults, are from the first generation of Northamptonshire kids in about two centuries to grow up with Red Kites as an everyday part of their lives. The kites seem attracted to playgrounds at break times, and there can be very few young adults who have grown up in these parts who do not know a Red Kite when they see one. We've stopped pointing them out to each other in the streets, though that did happen for quite a while, and now Red Kites are probably among the best-known and most-loved local birds. So much so that when the unitary authority of North Northamptonshire was set up alongside its sister authority of West Northamptonshire, we in the North were given a choice of three logos for our new council and two of them had Red Kites flying in our skies, and the better one of those two was chosen by popular demand. Here in North Northamptonshire we know our Red Kites, and we like them, and we have sufficient confidence in them and ourselves to stick them on our logo even though they were a rare sight when most of the councillors were born. That is progress.

Having lived two-thirds of my life in a world where Red Kites were rare and one travelled to special places to see them, I still thrill at every sighting, and I want to keep that feeling. Yes, they are much commoner now, but my baseline for Red Kites was set at an earlier time when they were rare, whereas my kids grew up with Red Kites increasing in numbers and today's children find them commonplace.

That is a shifting baseline concept which we usually encounter in a different way, where our memories are of more wildlife and we try to restore things to the status of our youths – not considering that our memories only take us a little way back along a long journey of decline and we ought, perhaps, to be more ambitious. But here in Northamptonshire our personal baselines for Red Kites have been a long line of zeroes, and that baseline triggers my thrill each time I see one now. There are very few conservation actions as successful as the measures taken for Red Kites – from zero to logo in 40 years!

Hedgehogs

Night is mammal time. Whereas birds are creatures of sound and colour, mammals are creatures of touch and smell. We feel more empathy with a Blackbird singing from a rooftop who greets the dawn, spends the day collecting food for its kids, and then goes to sleep as the light fades than we do for the snuffling rodents or flitting bats who emerge from their burrows or crevices to occupy the night. Well, I do anyway. And it's not as though our mammals are scary and command our respect; they are an odd bunch of rodents and misfits, and lack most of the top predators that once prowled our streets when they weren't streets but were still woods.

When I step into my garden at night, I occasionally see a bat fly past but that's about it. I've never seen anything as exciting as a Red Fox or a Badger in my street let alone my garden, but 20 years ago Hedgehogs were regular visitors. We would see them, find their droppings and sometimes hear grunting and snuffling as two Hedgehogs interacted amorously or aggressively – it was difficult to tell exactly what was going on in the darkness. But Hedgehogs have gone. I don't find them in the garden even though there are plenty of organic slugs for them to feast upon, and I don't see them dead on the roads nearby anymore. I recently saw a Hedgehog squashed on the road a couple of kilometres away and I envied that neighbourhood for having some Hedgehogs to run over by accident – ours are long gone.

Hedgehogs are amazing, aren't they? It would be difficult to dream up a Hedgehog if one didn't know that such creatures existed. The world has 17 species of hedgehog distributed across Europe, Asia, the Middle East and Africa, and hedgehogs have been on this planet for about 15 million years.

Hedgehogs are, of course, predators. I say 'of course' because every animal has to eat, and very few vertebrates are wholly vegetarian. Until recently, Hedgehogs were assigned to the order Insectivora, which gives a clue as to what they eat but not a completely accurate one as their diet includes molluscs (slugs and snails), annelids (a range of earthworm species) and I dare say some arachnids (spiders) and other invertebrate taxa. Most predators specialise but most also will have a go at eating anything that comes within their ambit, and for most that includes the occasional corpse too.

Many predators are termed 'vicious' predators by some person or another at some stage in their lives, but Hedgehogs have a pretty good press. That's partly down to Beatrix Potter's Mrs Tiggy-Winkle and is aided by the fact that Hedgehogs are declining dramatically in numbers, not just in my garden, and declining species tend to get their sins forgiven as a mark of sensitivity. We don't usually call Hedgehogs vicious predators because they are vicious to slugs and we'd quite like there to be fewer slugs in our gardens, on the whole, partly because slugs are slimy but all the more because slugs eat our vegetables. So the vicious slug-eating predator that is a Hedgehog is looked on favourably. We want it to be a vicious predator.

But the good-predator image can be turned around under exceptional circumstances, as we saw when Hedgehogs were deliberately and foolishly brought to the Outer Hebrides, probably to do a bit of good, it would have been thought, in somebody's garden. From there they got into the wild – which for ever has been a Hedgehog-free, Red Fox-free, Stoat-free and Badger-free zone. And then Hedgehogs started making inroads into the high densities of ground-nesting birds in that part of the world. Dunlin, Ringed Plover, Snipe and a range of other species experienced Hedgehogs gobbling up their eggs and causing declines in breeding success and then in population levels. On South Uist, Hedgehogs got a press that was distinctly different from that they enjoy in suburbia, they had been unmasked as the predators that they always are. Hedgehogs hadn't changed their diets as a result of their geography but we had seen more clearly a different aspect of the complexity of their dietary breadth. In my garden there are no Dunlin and lots of slugs eating my tomatoes. I'd like Hedgehogs back, please.

Grey Squirrels

I wouldn't say that a smile breaks across my face every time I see a Grey Squirrel in my garden, and that is not only because they sometimes dominate the bird feeders and exclude the birds I intend to eat the sunflower seeds I have provided. It's also because every time I see a Grey Squirrel I feel as though I am missing a Red Squirrel.

Grey Squirrels, a North American species, were tipped into the UK by landowners in the nineteenth century and have, through a combination of carrying diseases and being efficient competitors, caused a huge decline in Red Squirrels across most of England and Wales and some of Scotland and the island of Ireland too. Whereas once there were no Grey Squirrels and millions of Red Squirrels in the UK there are now around 150,000 Red Squirrels and 2.5 million Grey Squirrels – representing a massive displacement of a native species by a non-native.

In the 1960s there was a thing called the Tufty Club which told young children like me about how to cross the road safely on behalf of the Royal Society for the Prevention of Accidents. Tufty Fluffytail, a Red Squirrel, was the lead character and was accompanied by Policeman Badger, Willy Weasel, Minnie Mole and Mrs Owl. From the early 1960s, when I first started crossing roads on my own, to the early 1970s the Tufty Club grew to have over 2 million child members, and for a while I was one of them. A modern Tufty might have to be a Grey Squirrel, but would he or she talk to us in an American accent?

Grey Squirrels are loved by foreign visitors to St James's Park in central London, where they are tame and will take food from the fingers of enraptured children. There are no Red Squirrels to feed there but the children don't seem to care about that as they look round to their parents, beaming, as a squirrel approaches them, takes some food from their fingers and then retreats a little way to eat it.

To those children, and actually quite a lot of adults too, the power enshrined in the Treasury and the back view of 10 Downing Street at one end of the park, and in Buckingham Palace at the other, are as nothing compared to the thrill of a living creature approaching so close and taking a gift of food.

Grey Squirrels are known, by some, but not by many of the children in St James's Park, as tree-rats, which tells us quite a lot about what some think of them. They are blamed for damaging trees and gobbling up nestlings and eggs from birds' nests. Both happen. But the Red Squirrel can do no wrong and it is one of the poster-species of rural landowners who are keen to show that they care for wildlife. Funnily enough, the forebears of those rural landowners were responsible for the release of Grey Squirrels into the UK and were just as antagonistic to Red Squirrels before they became the loser in the struggle with Greys. There were squirrel destruction societies flourishing in the UK before Greys ever set paw here, and it is said that over 21,000 Reds were killed in the New Forest between 1880 and 1927, with over 2,000 in 1889 alone. These culls, of a native species, were motivated by Red Squirrels also being seen as pests of trees – and if we were to think that Reds wouldn't eat birds' eggs then that would be a mistake too. Squirrels are squirrels, they eat things.

I'd rather there were Reds running around in my neighbourhood but I'm not sure that the species-for-species replacement would make much difference to the ecology of my back garden, though I'd be keen to find out.

Cats

The most frequently sighted mammalian predator in our garden is the domestic cat. I'm neither a cat person nor a dog person, I'm not a great fan of enslaving animals as pets, but if I were forced to choose, then I might opt for cats because they behave a little more like their ancestral predatory forebears than do our ultra-bred wolf-substitutes called dogs. I don't own a cat, but cats visit our garden. They must be other people's cats, but whether the cats acknowledge their 'kept' status is unclear.

I liked one particular local cat which used our garden, one I dubbed The Little Predator. She got her name because she stalked our garden *as if* she were a killing machine rather than because I could lay any kills to her score. She was a tabby and her shoulder blades stuck up as she walked slowly, head low through the garden – just like those of a stalking lioness on television. The Little Predator replaced some of the predator pressure that is missing from most of our gardens. Her presence created a landscape of fear and the Blackbirds went bonkers as she paced slowly and menacingly near their fledged but naive chicks. I liked her because she looked real, dangerous and intent on death – she rewilded our garden a little. Of course, because she was a cat, she returned my respect with looks of complete disdain – she was quite cool.

Centuries ago, it is probable that a Brown Bear walked through the space now occupied by our garden, and some Grey Wolves, and Lynx and Pine Martens, Wildcats and Polecats. The past is a big place and things were different then. The UK lost its Brown Bears, not through carelessness but through intent, before the Normans arrived. The UK Brown Bear population of 7,000 years ago has been estimated at 13,000 individuals. The last Wolves probably went from England in around the year 1500. They are so long gone that we don't usually think of them as gone and it is difficult to imagine slotting them back into our modern landscapes.

The UK has a strange mix of mammalian predators – the big and bigger ones have gone, leaving us with Badger (approx. 12 kg), Otter (9 kg) and Red Fox (8 kg) as our largest numerous terrestrial predators, and the almost-extinct Wildcat (7 kg). They would once have been outranked by Brown Bear (250 kg), Wolf (40 kg) and Lynx (20 kg).

This unbalanced cast of mammalian predators must have big impacts on our overall ecology. Not only will some potential prey species (such as deer) benefit, but so do some smaller predators such as Red Foxes. So, if you believe that Badgers are wiping out Hedgehogs in the British countryside (which I don't, although the Badgers aren't exactly helping), then you might use that as an excuse to call for Badger culls, particularly if you are a dairy farmer, but you might be better off calling for the return of the Wolf, whose activities would cramp the style of Badgers and Red Foxes and maybe lead to a surge in Hedgehog numbers so that the slugs would suffer and I'd get more tomatoes each year. On the other hand, if you believe it's anything like as simple as that then you're wrong. This imbalance of nature thing is rather complicated.

I miss The Little Predator because I remember her stalking through our garden but I don't miss a large cat, the Lynx, in the same way because they are not within my UK memory, nor that of my parents or grandparents or even, really, within much accessible written history. Lynx are species of fable, and of faraway places, whereas domestic cats now pass through doing their best to impersonate top predators.

Falcons and flycatchers

I felt sheepish and chastened as my car was towed to Gells Garage after I foolishly attempted to drive it through a flood and the engine died. There was much sucking of teeth, blowing out of cheeks and looking at the ground, which seemed to mean 'You're a bit of a prat' – but then we spotted a Peregrine on the church spire. The mechanics who were going to try to resurrect my engine were also fans of the local Peregrines. We enthused about falcons, but their parting words were ominous: 'You might be looking for a new car, sir.'

Walking home through the churchyard, I looked at a gravestone for a perching Spotted Flycatcher. Of course, there wasn't one, because Spotted Flycatchers are in Africa in November, but they won't be there in high summer either, as they once were. I used to watch the Spotted Flycatchers here often – I'd take this route in preference to others just to see them as they sat erect, grey and with streaky breasts, looking for insects. Churchyards suit Spotted Flycatchers. The open nature of the habitat, with regular perches, is ideal for them. When an insect flies past, they can spot it, fly out and chase it, grab it and return to base to remove its wings and gobble it down. The Spotted Flys brought life, and admittedly death to insects, to the graveyard. I can see them in my mind's eye very clearly, just no longer in reality.

As well as in Raunds, I used to see Spotted Flycatchers in Higham Ferrers churchyard as I visited the farmers' market. As others passed with bags full of apples, cheeses or perhaps eating an ostrich-burger in a bap, I would watch the Spotted Flycatchers. They've gone now. The quiet, pretty churchyard of Achurch retained the flycatchers beyond when I last saw them in Higham Ferrers or Raunds, but they seem to have gone from there too.

Why have Spotted Flycatchers disappeared from my local churchyards? Is it a change of mowing regime or the use of new or different chemicals in churchyard management? Maybe, but I doubt it, as Spotted Flycatchers have declined not just in churchyards but in all habitats, and not just in Northamptonshire but across the UK, and not just in the UK but across Europe (but probably more in Western Europe than Eastern Europe), so we surely have to look for causes with a much greater reach than specific local ones. The bigger picture is that many avian summer visitors to Europe are declining in numbers, and that might suggest that the issues involved are climate change, overall reductions in insect abundance or problems on the wintering grounds rather than anything that is happening just locally in Northamptonshire's churchyards.

Life would be so much easier if the local losses of Spotted Flycatchers were due to local management. I could try to influence the management of my local churchyard by suggesting a change to mowing regimes or herbicide or pesticide use, or maybe a cull of Magpies or Grey Squirrels, but it is a different kettle of fish if we think that factors in Africa, maybe multiple factors, cause the loss of the sound of a Spotted Flycatcher's bill snapping on a fly in a Raunds churchyard. What are those factors, and how does one tackle them in countries as diverse geographically, politically and culturally as South Africa, Senegal and Kenya?

I looked back towards the church as I neared home and saw the Peregrine on the spire. Should I just feel happy that Peregrines have returned and sad that Spotted Flycatchers have gone? The Peregrine story can act as a beacon of hope, as this was a bird heading for extinction which was saved by near-global restrictions of the use of chemicals in farming. That took some doing, and maybe Spotted

Flycatchers can benefit from similar far-reaching international initiatives – but, if so, we haven't worked them out yet.

Happy endings are possible. The garage brought me back my car, fit and able to travel again, later that day. You just need to know what you are doing, and then do it, to turn disaster into triumph.

Birds and 'weeds'

I'm one of a few thousand volunteers who survey birds in a scheme called the Breeding Bird Survey (BBS). I visit a patch of farmland twice a year, in May and in June, and walk a prescribed 2 km route. On my early morning walks I note all the adult birds I see or hear and later enter the data on a computer and pat myself on the back. I've surveyed this 1 km National Grid square for 17 years and feel as though I know it quite well.

If you were to drive past my survey square, the sight of it from the road wouldn't invite you to explore it in more detail, but I am now very attached to it, even though it is an ordinary patch of arable farmland straddling the Northamptonshire/Cambridgeshire border. The crops grown here are mostly cereals and oilseed rape, although recently beans have partially replaced the rape. This is the factory floor of British arable farming where wheat is king and produces the money. If farmers could grow wheat every year they would, but if you try it then weeds, diseases and pests build up. Weeds, diseases and pests are different names for wildlife. For farmers, as for gardeners, there are only three types of wildlife, the good, the bad and the neutral, and it depends on your levels of tolerance where you draw the lines.

A grass grows along my survey route which I once looked up and found to be Slender Foxtail. It's an OK grass, but few would extol its beauty – it's a grass. To the farmer, once the Slender Foxtail gets into their field then it is a weed with black seedheads that pokes out above the wheat crop, called Black Grass. Slender Foxtail or Black Grass? Your choice of name will tell the world much about you.

Wheat is, biologically, a grass, and in searching for herbicides that get rid of most of the weeds of wheat, it is easier to get rid of flowering plants, because they differ physiologically more from wheat, than it is to get rid of fellow grasses. The main reason we see yellow fields of oilseed rape and green fields of beans around these parts is not that they are highly profitable crops but because they are a break crop of a flowering plant which is not a grass. Because oilseed rape is not a grass, you can use herbicides on it that you can't use on wheat, and that means you can get rid of more Black Grass. Growing rape or beans gives you an income from those crops but it also boosts your income from the following wheat crop being cleaner of weeds.

Our reliance on wheat (and barley) as the major arable crop over the last 50 years has led to a rather strange nomenclature for a group of formerly abundant plants which are now known as rare arable weeds. It's a fairly odd term but you can imagine how this came about. In the old days the likes of Pheasant's Eye, Shepherd's Needle, Prickly Poppy and many more, some with obviously agricultural names (Corn Gromwell, Corn Chamomile, Corn Spurrey, Corn Marigold, Corn Buttercup, Corncockle and Cornflower itself) were favoured by soil tillage and thrived, at the crop's expense. Little by little, changes in cultivation but mostly more efficient herbicides have shifted the balance of power to the farmer and, many would say, the consumer, and now we call them rare arable weeds.

I did record some birds on this BBS visit, but they have declined too. Before I started recording here, there must surely have been Turtle Doves and Corn Buntings in these fields, and a lot more Grey Partridges, but in my 17 years the numbers of Yellow Wagtails and Willow Warblers have declined almost to zero and even Robins seem to be on a downturn. Farmland generalists such as Carrion Crows and Woodpigeons are doing fine. It's a similar picture across the hundreds of BBS squares covering arable farmland: specialist farmland birds are declining and generalists are doing well. The names of many of the declining birds are related to farming too: Corn Bunting, Corncrake, Meadow Pipit.

Have we got agriculture right? Not for the Corn Bunting or the Harvest Mouse or the Cornflower, it seems, but we are told that our farmers are the best in the world and our food is of the highest environmental quality. There is something here that does not compute.

Big Garden Birdwatch

At the end of January each year the nation looks out of its windows to carry out the RSPB's Big Garden Birdwatch, spending an hour recording the birds of our gardens. This survey, firmly in the citizen-science camp, started in 1979, originally for children, and now over a million participants connect with garden birds.

My annual counts usually produce around a dozen common species, with House Sparrow being the most numerous, usually reaching double figures. Our garden is clearly fairly typical, as House Sparrow also tops the national BGBW abundance list and has done for the last 18 years. But House Sparrow numbers now are less than half their 1979 levels: the average of a few more than 4 House Sparrows per garden today used to be close to 10 per garden. House Sparrows have declined generally, not just in winter numbers, and not just in gardens. Research indicates that lack of winter food has been an important factor in the farmed environment (where are the stubble fields and hayricks of decades past?) and lack of insect food to feed to their chicks has been an issue in urban environments.

I feel lucky to have decent numbers of House Sparrows in my garden. They nest under my eaves and those of one my neighbours, so I get to watch them through the year, and with great interest. I tell myself that it's partly because I provide winter food for them (and a scruffy organic garden through the year) that I am so blessed, but there is a part of me that questions bird feeding on the scale that we carry it out in the UK.

Garden bird feeding is driven by two main motives, I guess. The first is wanting to help individual birds get through the winter's cold temperatures when natural food supplies are low and feeding is limited by short daylight hours. But we also have, you and I, a second, more selfish, motive for providing food for birds – we like seeing them, and if we spend a fortune on sunflower seeds then we will see them from our windows more often, in bigger numbers and with more species. Even if you were able to prove that my bird feeding did not increase survival or population levels of any bird species I would probably carry on feeding just to get good views of birds.

There is evidence that bird feeding is not wholly positive in impact. Greenfinches are susceptible to a protozoan parasite called *Trichomonas gallinae*, and feeders are prime spots for transferring the parasite. There is pretty strong evidence that recent declines in Greenfinch populations may be driven by trichomonosis. Perhaps my bird feeders are helping House Sparrow population levels but harming those of Greenfinches. Should I stop feeding? For now, I'm content to think that bird feeding does some good for individual birds, and for some species of birds, but I'll monitor the science closely.

However, I know that garden bird feeding is a tiny contribution to bird conservation (let alone to wildlife conservation), whereas some I have met regard feeding the birds as a major contribution and their only or main contribution. We apparently spend around £200 million a year on garden bird food. That figure is far more than the annual income of the RSPB. If we gave up feeding birds in our gardens, and instead donated the money directly to the experts in UK bird conservation, would we see more or fewer birds in our lives? Would House Sparrow numbers plummet, or are we fooling ourselves that we are really helping them? Would Greenfinch numbers bounce back as the lack of feeders enabled them socially to distance and reduce their chance of catching trichomonosis? Would our larger, better-managed wildlife reserves be richer in wildlife at the expense of seeing fewer birds out of our kitchen windows? Discuss.

House Mice

I like wildlife, but there is a limit. A few winters ago, I kept wondering whether I had seen something move out of the corner of my eye and I remembered what that might mean from my youth – House Mice. One night there was a rustling noise under the bed too which stopped every time I moved. And then I saw a mouse scurry across the tiled kitchen floor. It was very cute. I liked it. We can cope with a few mice running around, I thought to myself.

Then they started turning up in the living room and watching television with us, and at one time I could see two mice in different corners of the room. Was this getting out of hand? Our children thought so when they came home and a mood was building that the mice had to go and that Dad/Husband/Mark was being a wimp.

Whilst still contemplating the next move, one evening a very cute House Mouse scurried near my chair and I took a paper bag out of the bin and let it investigate it before scooping it up and crying 'I've got one!' But what to do with it? I decided that I'd release the mouse outside so I very gently tipped it out onto the pavement. I can't say it ran between my legs and back through the front door, but it did run quickly past me in the direction of the house and it might well have been that very House Mouse who was in the corner of the room when I got back inside.

Mouse removal, it was decided, was very much a man's job, which seemed to mean me. We had a couple of mouse traps that hadn't been used for about 15 years but seemed perfectly serviceable when I trapped my finger in one of them and we added to their number with four more purchased specially. So these six break-back mouse traps were baited with peanut butter and put at strategic locations where examination showed there were some mouse droppings, and left unset for a couple of nights. Then the evening came for them to be set before I went to bed, and as I headed up the stairs I heard one trap go off. 'Rats!' I thought, 'I must have set the trap too sensitively.' But there was a dead House Mouse in one of the traps. Its corpse went into the garden, and the trap was reset.

In the morning all six traps had dead mice in them. And over the next few nights we reached double figures – and we haven't seen a mouse or its droppings since. Job done.

I tell this story because it is a handy, personal, and true example of how our attitude to wildlife can vary with the circumstances. Wildlife is great, except when it is bothering me, and bothering comes in many shapes and sizes.

We could have used non-lethal means such as putting every item of food in sealed containers everywhere in the kitchen – that would have been quite a big job. We could have trapped them and released them, but my experience with the paper bag wasn't encouraging. I have mentioned this to friends and many of them, it tends to be the women, suggest trapping them and releasing them up the road so that they don't come back to my house – that doesn't seem very neighbourly to me – these are House Mice, they live in houses. Should I post them through my near-neighbours' letterboxes? Or is the trans-location scheme simply to wash one's hands of them and assume they have found a new home and are being cherished elsewhere? We could have got a cat, but there is a history of biological control attempts that have gone wrong when the introduced predator eats the wrong species or proves to be totally useless. And anyway, cats are more expensive than mouse traps and offloading the moral responsibility onto a feline doesn't seem right to me. So I took action to protect my home from the House Mice and I won.

We usually win in conflicts with wildlife.

Stanwick Lakes

Many birders have a local patch – an area they visit regularly and with which they feel an affinity through knowing it and its birds well. My local patch is Stanwick Lakes, formerly gravel pits, now a country park with a visitor centre, playground, car park, tracks and cycleways. It's where I do much of my birdwatching. I can just remember it as an area of wet grass fields before gravel extraction.

I am very much a spring birder. I love the spring with its ever-improving weather, bird song and the return of migrant birds. It is at Stanwick Lakes that I usually see my first Swallows and Swifts and hear my first Chiffchaffs and Cuckoos of the year. I'm a frequent visitor as spring uncoils.

Nine warbler species occur very regularly at Stanwick Lakes. Cetti's Warblers are present all year and sing on any sunny day in any month. Chiffchaffs and Blackcaps are present through the year but only in small numbers in winter. Chiffchaffs start singing in mid-March when most arrive back in the UK from wintering in southern Europe and North Africa. Blackcaps are slightly later, towards the end of March – usually just before the Willow Warblers come back from sub-Saharan Africa and fill this patch of North Northamptonshire with their liquid cascading notes. These are followed by Sedge Warblers, Whitethroats, Reed Warblers and Garden Warblers, and the last of the lot is the Lesser Whitethroat.

The roll-out of warbler song is predictable but not wholly so, and that variable regularity brings me back day after day in spring after spring. There has never been a year when Whitethroats have arrived before Chiffchaffs, and I am confident there never will be, but I want to experience how spring unfolds each year.

My relationship with Stanwick Lakes is rooted in its birds, and goes back over 30 years. I used to see Turtle Doves here every May/June through the 1990s and into the early 2000s, but my last record was in 2008. That reflects their decline locally, where they are now few and far between, and nationally, where their numbers have fallen more than those of almost any other UK breeding bird – by 95% in 25 years.

It's not all loss – there are some gains as well. The most obvious newcomers are white egrets of three species: Little, Great and Cattle. When I first started birding in Northamptonshire, to see any of these would be a mega-event. My first Stanwick Lakes sighting of Little Egret was in 2005, of Great Egret in 2014 and of Cattle Egret in 2018. Suddenly, seeing all three species in the same visit is quite normal.

In my leisure time at Stanwick Lakes, I am very much a birder and not a naturalist. If you could tune in to the internal monologue in my head as I walk these familiar paths it would be very bird-dominated, with thoughts of what that distant flying bird is and what produced that burst of song, as well as an awareness of whether I've seen the expected species so far on this visit and what I usually see, or might see, just around the bend in the path. I pay attention to the birds I see, I watch what they are doing, and the scientist in me notices unusual or interesting behaviours.

Although birds are my thing, I can tell you something about other wildlife here, like where I've sighted Otters, where the Grass Snake might be seen, and where the patches of Cuckooflowers grow where one can watch Orange Tip butterflies laying their single conical orange eggs under the flowers. But I'll admit to being a poor botanist and entomologist – interested but not very good. I'd like to do my regular walk with experts in other taxa – they would open my eyes to many things that I walk past and, in my ignorance, ignore.

This patch is an important window into the natural world for me, and I feel the changing temperatures of the seasons, notice the spring buds and autumn colours, and am aware of the emergence of dragonflies and butterflies and the flowering of the commoner plants. But when I wake and think of heading down to Stanwick Lakes my motivation is probably 80% 'I wonder what birds I'll see?' 15% 'It'll be good to have a walk', 5% 'what other wildlife will I encounter?'

Stanwick Lakes has hundreds of thousands of visitors each year. Most of these are using the café and/or the playground. The early-morning visitors are a mixture of a few birders, some dog walkers and many joggers and cyclists. Only a handful will be sad about the loss of Turtle Doves and very few will realise that three species of white egret occur here and are recent arrivals. Most of the joggers seem to have music playing in their ears as they run and so probably won't notice which warblers sing or whether their songs cease. Despite this site having all sorts of official designations because of its importance for wildlife, most visitors are blind to that importance and fairly blind to the wildlife as well. That's entirely understandable, and it's fine by me (my grandson loves the playground too), but it's worth remembering that one person's window onto the natural world is another person's car park near a café. We don't all see the world the same way.

Swifts

Raunds has two vaguely famous former residents: Ada Salter and David Frost. Salter, born 1866, grew up in Raunds. She became a social and environmental campaigner and was the first woman mayor of London and the first Labour woman mayor in the UK. Frost, born 1939, was a journalist, comedian, writer and television host whose father was a Methodist preacher. The Raunds Wesleyan Methodist Church was established in Brook Street in 1812 and rebuilt in 1873/74. At their different times, Ada Salter and David Frost both sat in its pews and maybe thought of how the world could be a better place.

The 13 trustees of the chapel when it reopened in 1874 were all men, some with surnames still common in Raunds, and their occupations tell us something about nineteenth-century Raunds. They were four farmers, an accountant, a baker, a shopkeeper and six others associated with leather: a shoe manufacturer, a shoe agent, a saddler, a currier (of leather) and two clickers. What's a clicker? A clicker cuts the upper for a shoe or boot from a sheet of leather. You depend on your clickers to make the most of your leather with least wastage and also to produce the best shoes by ensuring the leather can stretch with use by taking note of the equivalent in the leather of the grain in wood. Their cutting of leather, with hand-knives, leads to a constant clicking. The first two owners of my house on Lawson Street were clickers.

I cut through the Methodist graveyard on my way to and from the post office if I take the shortest route. On the way home, particularly on a sunny day, there is the temptation to stop, perch on a stone memorial and watch the butterflies in the grass between the gravestones: Meadow Browns, Ringlets, Gatekeepers and Common Blues, with Painted Ladies in those years when they are common.

Above my head there will be Red Kites, which were not here in the times of Salter or Frost, but also Swifts, which certainly were. The Swifts arrive in late April and are a constant screaming presence as May turns to June, and June to July – but in early August their numbers decline and a September Swift is a rarity. Swifts are with us for a mere three and a half months, but those reared in Raunds are likely to return here to breed after they have spent their first two years in constant flight, sleeping on the wing (can you imagine it?) and catching insects that you and I have never seen or heard of over the tropical forests of the Democratic Republic of the Congo and Mozambique.

Swifts nest in crevices in our buildings, which raises the obvious question, 'Where did they nest before there were buildings?' The answer is that they nested in holes in old trees in old woodlands, and a few still do. I've been shown Swifts nesting in the most ancient Scots Pines in Abernethy Forest, in former Great Spotted Woodpecker nest holes. Both there and above the busy Brook Street of Raunds in June and July Swifts feed on the wing in summer. One of the nesting sites for Swifts is the Methodist chapel.

I wonder whether they nested in this same building 150 years ago when Ada Salter attended – maybe not, as the building was new then and maybe it had not quite developed the cracks and apertures which a few pairs of Swifts use now, but surely David Frost would have been able to see Swifts going in and out of the chapel roof and maybe heard these birds, sometimes referred to as Devil Birds, screaming outside as his father preached to the congregation. Maybe?

The time has come to repair the Methodist chapel as a big lump of masonry fell from the roof recently. So, the roof is going to be repaired. A local naturalist was quick off the mark to alert the chapel to the Swifts' nesting sites and succeeded in getting the necessary work delayed until after the nesting season, and we are now hopeful that the refurbishment will include a Swift nest box.

Flying Ant Day

Few days of the year capture the differing attitudes to wildlife among our population better than the reaction to 'Flying Ant Day'. Flying Ant Day is a natural phenomenon to be noted and celebrated – it's one of the few times when wildlife bursts forth in impressive numbers all at once, presumably as it always has on a day towards the end of July or in August each year. On this day, which does vary from year to year, and from place to place, winged ants take to the air to find mates. In practice, there isn't one single day and flying ants may be found on many summer days – but the combination of high temperatures and low wind speeds on fine days are not so common in an English summer that clumping of events spatially doesn't occur. I imagine the ants in a particular nest being more or less ready but then being a bit picky about when they will take to the air. All you need is a period of less-than-ideal weather and the ants are queued up underground waiting for the right conditions to take to the air – and then, after a period of unfavourable weather all that pent-up lust can no longer be contained. Well, that's what I imagine is happening in the nest of the small unobtrusive black ants in my garden – a species known, appropriately enough, as the Black Garden Ant.

The ant nest is under the garden shed, somewhere. I just see a few ants coming and going in summer, disappearing down cracks in the garden path alongside the shed. But on one day of the year streams of silver-winged ants emerge above ground, run around momentarily on the path and then take to the air. I love to see it. I notice it most years, but it is sufficiently unpredictable to take me by surprise each year. And sometimes I miss it simply because I've gone out for the day and it's all over by the time I return. But when I see it, I love it. It's as though a tap of insect abundance has been turned on and is gushing – gushing from my garden.

Just for a while everything is different. The path has scores of black ants with long silvery wings sparkling in the sun and the air above the path sparkles with their flight. Higher up I cannot see the ants themselves but the feeding behaviour of the local birds shows where the ants have gone and many of them must be very high in the air, hundreds of metres, according to the birds. There are Starlings trying to be Swallows capturing the ants at 6 metres, and above them there are Black-headed Gulls pretending to be Nightjars, feeding on the flying ants at 30 metres, and above them there are Swifts, knowing full well that they are Swifts, feeding on the flying ants much higher still. It's a short-lived feeding bonanza and it looks as though it's not just my garden where it's happening because there is an awful lot of frenzied feeding going on at the same time.

But when this happens, I know I will see coverage in the media, not usually locally but often from London, where people are going wild about the 'invasion' of 'huge' ants which 'don't pose much danger' to people (do they pose any danger at all?). The *Daily Mirror* once listed six ways to get rid of flying ants: spray them with dishwashing soap, catch them with sticky tape, attack them with toxic artificial sweeteners or insecticides, place tin cans over the anthill and pour boiling water down the ant hole. Well, what a nation of animal lovers we are! One year, the emergence of the 'huge' ants (I think tiny ants would be more accurate) coincided with Wimbledon and there were images on TV of stars of the sporting world who can cope with the pressure of being two break points down in front of an audience of millions but not with some ants flying past. To be fair, the air full of ants is distracting, but it isn't an assault on humanity, it's a short-lived inconvenience for some, but not many, and it ought to be a joy to many more. Life bursts forth every year!

Blackberries

There are two blackberry patches in our medium-sized garden and they provide enough fruit for many blackberry and apple pies and crumbles. They differ in several ways, perhaps reflecting the complex biology of the Bramble which exists as hundreds of genetically different microspecies. Patch 1 produces many small, hard, slightly bitter blackberries. They do well when cooked but I'm not tempted to eat many of them as I pick.

Patch 2 is more recumbent, has thicker, stiffer stems but delivers big, juicy, tasty blackberries. Whereas on Blackberry Patch 1 there are green, red and black blackberries sitting next to each other, on Patch 2 if one blackberry has made it to the tempting purplish black of ripeness then so have almost all of the others. This makes picking much easier too. The main snag with Patch 2, if you are 63 years of age and 190 cm tall, is that these are low-hanging fruit – so low that there is a lot of bending to be done, and that is a pain both metaphorically and literally.

As a child, I went blackberrying on family trips from Bristol into south Gloucestershire to fill Tupperware containers. My maternal grandmother used to sing out 'My patch, my patch, nobody come to my patch', words and a tune she had learned in south Wales in the last years of the nineteenth century.

Nowadays, we make raids into nearby Cambridgeshire to a green lane where the hedge, for about 100 metres, is laden with blackberries. It's off the road, down a track and there is easy parking near a house. We never pick the blackberries in front of the house, tempting though they are, but walk a way down the green lane where we regard the blackberries as fair game for all. Forty minutes later we have two full litre-sized containers and purple-stained and sticky fingers and a few scratches, but it's well worth it. A few sunny September days each year should include spending an hour working your way along a hedgerow and picking these fruits.

I can see that others pick blackberries where I do because the ground is somewhat trampled in exactly the places where I stand to make the most of my height and reach to harvest the more distant fruits. How many people have stood on this very spot and reached out for ripe blackberries, and how many have felt the sharp pain of a thorn wound and the lesser pain when a ripe blackberry that you intended to pick falls to the ground as you pluck one of its neighbours?

Along with sloes and elderflowers, and mushrooms if you know what you are doing, they form the easiest aspects of wildlife to harvest ourselves, directly, and assuage some of the gatherer aspect of the hunter–gatherer in us all. Hunting for meat takes more time, more kit, more effort and involves a lot of other issues such as land ownership and shooting rights.

Blackberrying always feels worthwhile and I never feel that I am depleting a natural resource unduly, as when I leave the hedgerow most of the black blackberries, and all of the red and green ones, are still there for a host of insects, birds and mammals to consume. By its very nature a bramble patch is a self-defended pile of thorns, and picking blackberries entails taking blackberries from the edge of the patch without getting too scratched as you do it. There are always blackberries out of reach and they always look like the best ones. This feels like a very sustainable harvest.

Pheasants

Raunds is a rural town, and I am reminded of the nearby countryside by the sounds that reach my ears when I am in my garden, embedded geographically, even if less and less culturally, in the countryside. For a day or two each summer I hear the combine harvester as harvest progresses, and in winter the explosions of unseen gas guns, to scare Woodpigeons from the crops, reach me. On a spring morning the local Blackbirds dominate my dawn chorus and, rarely now, I sometimes hear a distant Cuckoo – but much more often the crow of a cock Pheasant. The countryside is just a stroll away, minutes away by foot but less than a second as the sound travels.

Pheasants don't come to my garden, just their raucous cries, but in autumn there will be Pheasants for sale in local butchers' shops and I'll sometimes encounter Pheasant-shooting parties in the fields. The number of Pheasants locally shot are fairly modest – a few farmers clubbing together with their friends to release some birds and have a few days' shooting as a part of the autumn scene, just as blackberrying is for me and picking mushrooms is for others. But Pheasant shooting can be big business, with a day at a prestigious and well-stocked shoot costing over £1,000.

Recreational Pheasant shooting has grown massively. Today, 51 million Pheasants are reared and released each year in the UK, and 13 million are shot and killed. Back in the 1970s the number of released birds was around 4 million. That scale of release, of non-native gamebirds, is quite staggering – and in late summer the combined weight of those Pheasants, plus 11 million Red-legged Partridges, has been calculated to be greater than that of all native UK birds put together. Until recently, no-one has really considered the ecological impact of this rear-and-release form of shooting. Pheasants are omnivores and gobble up plants, seeds, insects and even some reptiles and small mammals. There are some worrying known impacts and a long list of worrying potential impacts.

I sometimes hear Pheasants described as invasive non-native species, but we should count our blessings that they are not. Only around 4 million Pheasants, make it through the shooting season and nest as 'wild' birds in our woods and hedgerows. Although 13 million Pheasants are shot, the other 38 million released birds die from disease, starvation, being run over on roads and from being killed by predators, most notably we think by Red Foxes. These headline numbers tell us that Pheasants are not invasive, they are quite the opposite, they don't survive in our countryside very well at all, which is why another 51 million birds need to be reared in captivity and released next year.

When I sit in my garden and hear a Pheasant call, I think of these issues. Ceasing Pheasant releases would be a fascinating experiment. I think we'd see some welcome and notable increases in the numbers of invertebrates and maybe of threatened reptiles such as lizards and snakes as well as a drop in Red Fox numbers, which might have knock-on impacts on other species. But, instead, we are at the sharp end of a non-planned, non-agreed and non-monitored experiment of the opposite sort where we have drifted into a massive ecological assault being perpetrated on our wildlife. When I hear a Pheasant crow, I can only salute him as a survivor in a foreign land shouting to the world, including me in my urban garden, that he has dodged the guns, dodged the Red Foxes, dodged the road traffic for another year, and he is ready to breed now that spring is here.

Bathroom wildlife

Our bathroom looks onto our back garden and the roof of an Ivy-covered shed. The Ivy, which is old, thick and tall, growing to a metre or so above the roof of the shed, is used by wildlife throughout the year. For a start, it is wild life itself, a glorious native plant which flowers in late summer and through the autumn and has its berries in late autumn which last through the winter. The smell of flowering Ivy is not my favourite aroma but the plant's attractiveness to a wide range of animals makes it a great asset.

It's always worth looking out at the Ivy on a visit to the bathroom. In spring there may be a Brimstone, perhaps the first of the year in March, sunning itself, in autumn Red Admirals feed on the Ivy flowers, with perhaps over a dozen in view in the September sun, and at stages in between Holly Blues flit. Ivy Bees, a recent colonist of the UK, feed on Ivy flowers in late autumn, but you can't be sure to identify them from the bathroom window – you have to get a much closer look, particularly as they are mixed in with wasps and hoverflies.

Woodpigeons feed on Ivy berries in autumn. I watch them as I shave. They get into amazing contortions, sometimes hanging almost upside down with a wing spread to aid their balance, and they seem keen on particular clumps of berries, reaching past more accessible ones to eat the ones they really, really want.

Lying in the bath in summer, with the window open, I can hear birds and insects as well as sounds from nearby houses and gardens – lawns being mown, nails being hammered, barbecues being had. There is some traffic and aircraft noise, an ice-cream van's siren, and sometimes loud music from a particular nearby house, and a surprising amount of clucking of hens and barking of dogs. Three times a day, the sound of play time from a primary school carries in our direction. The human sounds mingle with the remnants of the natural sounds.

But to enjoy the sounds of the birds and the bees from a horizontal position I need to run the bath, and that necessitates seeing whether anyone, any creature, is already occupying it. There is quite likely to be a spider in the bath and sometimes a silverfish – but never a bird or a butterfly.

Silverfish are small, wingless, largely nocturnal insects which are sometimes regarded as pests because they eat a wide variety of starches and include paper such as books in their diet. Now I come to think of it, I don't recall seeing a silverfish in the bath for quite some time, maybe years. I'd like one to pop up so that I can find out which of the three UK species it is.

Spiders in the bath are commonplace. They don't freak me out – they interest me. As I understand it, they get into my bath by seeking water and not being able to escape, so I rescue them and release them on the bathroom floor. As best I can tell, they are mostly Giant House Spiders, which spin sheet-like webs to catch their prey. There are, I notice, spider webs in my house but I rarely see flies caught in them. Is this because the spiders get to them quickly or because my home is not a great place to make a living as a spider? I'm really not sure, and I'm really not sure about much when it comes to spiders.

My ignorance of spiders is huge – I'll willingly admit it. There are, in my defence, a lot of spiders in the UK – some 640 species – and few accessible books about them, but spiders are important predators of insects right across the world and they live in my bathroom too. On farmland there are something like 800,000 spiders per hectare, which equates in my calculations to nearly 2,000 spiders in the area of my bathroom – so as far as I can make out that bathroom has a deficit of getting on for 2,000 spiders. Or do I just need to look harder?

Easter

Easter Sunday is the Sunday after the first full moon on or after 21 March – so it moves around the calendar within a period of about a month. During that period the Pasqueflower blooms in a small number of sites, mostly in southern England. Pasque means related to Easter.

The Pasqueflower is such an attractive flower, with its purple sepals (Plantlife is quite fierce about them not being petals but the difference is lost on this ornithologist) and bright yellow anthers that even I can identify the species, and am drawn to have an annual look at it. Although Easter jiggles around in the calendar, and Pasqueflowers take more notice of the recent weather than the historical date of a crucifixion to flower, the timing is broadly predictable and the location is entirely predictable.

Pasqueflowers are found at a few dozen UK sites and pretty much exclusively on areas of short, chalk grassland. One of the top five sites, Barnack Hills and Holes, is on the far side of the A1, in Cambridgeshire just south of Stamford at a comfortable 40 minutes distance from home.

Since we are talking Cambridgeshire, the hills hardly trouble the contours of the Ordnance Survey map and the difference in elevation from the bottom of the holes to the top of the hills is only about 5 metres. On arrival at the edge of the small site, of 23 ha, I park and glance at the map of the site, put there by Natural England as this is a National Nature Reserve, but I know where to go. I head to the centre of the site, through another gate, and to where certain areas are cordoned off so that the plants are protected from trampling.

This is a transformed landscape. Stone was quarried here by the Romans and in later centuries, and about a thousand years ago rock was taken from here for the two great local cathedrals of Peterborough and Ely. But the children gleefully running or rolling down the steep slopes are not bothered by any of that.

If my timing is perfect then the Pasqueflowers look superb – there are hundreds of them in numerous 'flocks' spread over the slopes inside the fenced areas (with one or two outliers which allow particularly close views), but even if I've jumped the gun a little and they are not in full bloom, or missed the main flowering and am looking at seedheads, it is still worth the visit to see a beautiful once-common and now very localised plant, which survives here through a mixture of chance and design.

This site survived because the undulating nature of the post-quarry landscape does not make the site conducive to the plough or to building, so it has stayed as an area of grazing. In more recent years Barnack was notified and designated as a Site of Special Scientific Interest (SSSI) and as a Special Area of Conservation (SAC), both labels designed to protect the wildlife interest – and they have worked.

Barnack is not just a Pasqueflower site, for it is rich in rare orchids later in summer and is a great place to find glow worms on a summer evening, but the draw of the place for me is certainly the Pasque-flowers. They seem like survivors to me, like Hen Harriers and Turtle Doves, survivors from a past abundance of which we can still occasionally encounter fragmentary reminders. We must have lost 99% of our Pasqueflowers in the last century, and now 99% of the survivors occur on a handful of sites of which Barnack is one. It's an Easter pilgrimage to honour the living and the dead.

Heading home, I pass a signpost to Helpston, home of the Northamptonshire poet John Clare, who knew the Pasqueflower and on 25 March 1825, a week before Good Friday, wrote that he 'coud almost fancy that this blue anenonie sprang from the blood or dust of the romans for ... it grows on the roman bank', and noted that the Pasqueflower 'did grow in great plenty but the plough that destroyer of wild flowers has rooted it out of its long inherited dwelling'. Clare mourned the loss of Pasqueflower whilst celebrating their survival in places. I do the same, following much further loss and their survival in many fewer places.

Daisies

I'm very fond of the Daisies that grow in my lawn. Actually, 'lawn' gives the wrong impression. It's grass and flowers, not what most people would call a lawn, or if it is a lawn then it would be described as a very weedy lawn.

I've been letting some grass, by the fence, grow uncut for most of the summer for many years – it's there that the Large Skippers, Meadow Browns, Ringlets and Gatekeepers are seen in summer, but there is still grass to be cut because variety is good and some of that grass forms the access route to washing lines, tomatoes and other places of interest to me. And so I cut the grass now and again.

It is a source of amusement to my nearest and dearest, and perhaps a small amount of irritation, that I often spare any patches of Daisies from the mowing, and a good few Dandelions too. This is on the grounds that I don't want to decapitate them on a sunny day when I can see bees and other insects perching on them to suck up nectar. And, let it be clearly stated, Daisies are pretty, and I like them.

This means that our grass tends to have a tripartite character in summer: there are the unmown areas which have metre-high grass in late summer, there are the most recently mown short-grass areas, and embedded within those there are medium-length Daisy and/or Dandelion patches. The medium-length patches move around the short-cut area through the summer. I make these decisions partly on the grounds of wildlife benefiting, but also on the personal view that aesthetically a mown area with patches of Daisies and Dandelions looks nicer than a uniformly cut area of grass.

Even the RHS leans slightly my way on its website, stating that 'A sprinkling of wildflowers in a lawn can be a joyous thing for gardeners and wildlife. However, where a gardener chooses to create a more traditional green swathe, some control of the plants that are not grass may be needed. This can be done by good cultivation and by digging out, but lawn weedkillers are widely available too.' Traditional? Pah!

As I have often confessed, I am no botanist, but I do like the common or garden Daisy. I hadn't realised that they get their name from 'Day's eye' but I certainly knew that they closed up at night and reopened in the day. I often go to look at my Daisies in the evening, but no, I don't wish them goodnight or talk to them at all, but then, neither do I chop them off for a false sense of tidiness – so I think that my Daisies fully understand that I am on their side.

Reflection 1

No day passes without me seeing wildlife. We are steeped in it, surrounded by it – and that is as true in my small Northampton-shire town as almost everywhere else on Earth. I cannot look out of a window, spend time in the garden, walk through the streets or stroll in the countryside without encountering wildlife. Admittedly, my daily life does not include coral reefs, wolves or rainforests but it does include the likes of Herb-Robert, House Sparrows and Black Garden Ants. There is enough wildlife within 20 metres of my home to fascinate and delight me. Our world is far from a science-fiction nightmare of metal and glass without a plant in sight. Wildlife is not elsewhere, it is everywhere.

For me, plants and animals have been a source of fascination from childhood to grandparenthood. As a child I was simply amazed that creatures as weird-looking as spiders and hedgehogs could exist. That fascination was the starting point for becoming a scientist studying animal behaviour and then a conservationist trying to give wildlife a more secure future. I've never lost that heartfelt love of the wildlife around me, and that's why I still pay attention to it in my house, in my garden, in my street and in the nearby countryside and wildlife reserves. Wildlife conservation is my profession but wildlife is my passion and my life has wildlife written through it, and my friends tend to share a very similar outlook. We are highly engaged with wildlife and sympathetic to it. Our conversations are peppered with mentions of wildlife as well as news of family, work and holidays.

If my friends ruled the world, then they'd give wildlife conserva-tion a much higher priority in public life and I would cheer them on even though we all have slightly different takes on the subject. I would be amazed if any readers of this book, even amongst my friends, have exactly the same relationship with wildlife as I do. Some detest cats, some will agree with the RHS view of a good lawn and many, I suspect, will think less of me for killing some House Mice. That's OK, even amongst friends there are differences of opinion, sometimes serious ones, on the things that we care about – from music to politics and from child-rearing to food.

If I were to stroll through Raunds again, and instead of looking for Herb-Robert ask those I met whether they liked wildlife, most would answer that they did, with differing degrees of enthusiasm. No-one is likely to say they hate wildlife. And yet there is hardly a species I have mentioned so far about which there is not some controversy. The Herb-Robert I love is considered by some to be a weed, the Peregrine I salute will sometimes take someone's racing pigeons, and the Black Garden Ants which I watch in awe as they take to the skies simply annoy others. The Red Squirrels we now love were once much-hated, as are the Grey Squirrels that now delight visitors to London parks. We got rid of top predators many years ago and now we moan when cats walk through our gardens. Our society's collective relationship with wildlife is theoretically positive but has quite a lot of grit in it.

For all the people who root for wildlife or want to do some of it some harm, there are scores who basically don't notice it is there. Most of our fellow citizens don't have time to be bothered by plants or insects because they are busy with work, loved ones and the cost of living, with getting the car serviced, doing the shopping and attending to domestic chores. Rather than being a society of wildlife lovers or wildlife haters I suspect our overall relationship with wildlife is one of indifference. Let us be clear, most people in your street and of your acquaintance are pretty wildlife-unaware.

My daily encounters with wildlife are precious to me; they inspire me and are my personal glimpses into the natural world. You will have similar and different encounters and thoughts of your own. However, to judge our society's overall relationship with wildlife we should look at our impacts. How has wildlife fared in the UK? That is the subject of Chapter 2.

CHAPTER 2

The state of wildlife in the UK

My relationship with wildlife is informed by my lived experience of it. When I see a Red Kite overhead in Raunds I am thrilled by the bird itself, its mastery of the air, its plumage and its call, and anyone else living here can get the same experience just by looking up and opening their eyes, their ears and their mind to the wonder of it. But my thrill at seeing or hearing a Red Kite has baggage attached to it, and in this case the baggage makes me appreciate the encounter all the more because I know, from living in this town for 35 years, that what I am seeing would not have been so easy to experience even 10 years ago. I can also add the baggage that people I know, and to a small extent my own work, helped to bring that wonderful bird to the improved status it now occupies. My Red Kite sightings are made all the more enjoyable by the baggage attached to the sighting, but it works the other way too.

When I visit the silent Nightingale wood just for a walk, or to look for butterflies in summer, it's a great place and I may well encounter wildlife that pleases me, but I can't completely shake off thoughts of the missing Nightingales. I cannot hear their absence but I know it, and I know it from personal experience. That's the baggage attached to this place. If I hear a Nightingale elsewhere, I am momentarily carried back to this wood, the wood where Nightingales have been most important to me, and as a conservationist I feel remorse that we haven't ensured that more woods keep their Nightingales.

Generally, I think this baggage is a good thing. If you are interested in wildlife then your sightings are likely to compel you to find out more about what you see, and with that knowledge comes context – which influences how you feel. An informed relationship with wildlife must be a good thing but it sometimes feels like a burden, which is why I call it baggage here. Like me, you may be pleased to watch a Grey Squirrel visit your garden, but the baggage of knowing it is the cause of the loss of the Red Squirrel from much of the UK does detract from the pleasure of the sighting. Sometimes it would be nice to shed the baggage, but the privilege of doing that is largely restricted to the very young, the visitor to a previously unvisited continent or an alien lifeform visiting Earth for the first time. And in all cases, the baggage is likely to accumulate over time.

The state of wildlife, species by species and overall, affects how we think of it, and so it should. We feel differently, individually and collectively, about rare species and common species, and about species that are increasing in numbers and those which are decreasing. This is somewhat analogous to the fact that we feel a bit differently towards rich people and poor people, and ill people and healthy people – yes, they are all people, but our sympathies go to the underdogs. And in the case of human society those sympathies lead (or should lead) to very large amounts of the wealth of nations flowing in particular ways through taxation to social and health services – those are societal responses to the collective plights of many individuals. When we consider the wildlife around us, we appreciate, with a bit of thought, that there will be some species that are numerous and others that are rare, but we don't expect to find that all species are increasing in numbers or that all are decreasing. If many species are getting rarer then, quite rightly, we question why that is happening, and we tend to be concerned about it, especially if we notice that all those species declining in one place whether locally or nationally, are doing fine in another locality or in another country. If we discover that all the wildlife around us is stable in numbers and geographic distribution then that is a very different situation from if we find that the UK's wildlife is in decline. And in the case of widespread declines, we might want to intervene with measures which redress the balance – but unless we know what is happening, we can't even consider such responses.

So how do we get a rounded and accurate context for the state of the UK's wildlife that can inform our future relationship with it? If I based my assessment on my own local experience it would be a pretty unreliable picture, largely because my observations are restricted to one landlocked county of southern England, because my memories and records would be a bit suspect, but also because I'm good on birds and very poor on many other taxa. But, added to others' observations, and with a bit of rigour and structure mixed into the collection and collation of those records, our experiences become a much more powerful account of what's what. They become data and not just sightings. They will give us a good picture of the state of wildlife in the UK.

What counts?

You can tell a lot about human existence by what we count, how accurately we count it and how often we report the results of all that counting. And if you look at our world through that lens then you will unavoidably come to the conclusion that money not only talks but also counts, and that the state of wildlife around us isn't treated as being of much import.

Money is a useful construct to help us exchange goods and services, and is sometimes used as a proxy for happiness, but the link is very unreliable. And yet we measure it with great care and precision. I get a text from my bank every morning telling me how much money I have in my account – to the penny. Right now, I can look up the monetary value of any publicly quoted shares and see their value change in real time, and various stock market indices changing with them, all the time, as I watch. The exchange rate of the pound with other currencies changes in real time. The inflation rate is measured and reported upon, as also is the size of the economy and the national debt.

Much public debate is framed in terms of money rather than in terms of life, liberty and the pursuit of happiness, none of which, the last time I looked, is easily quantified in cash terms. To paraphrase the incomparable Douglas Adams, writing in one of the finest environ- mental books of all time (*The Hitchhiker's Guide to the Galaxy*) 'Most people were unhappy for pretty much of the time. Many solutions were suggested, but most were largely concerned with the movement of small green pieces of paper, which was odd because on the whole it wasn't the small green pieces of paper that were unhappy.'

Why don't we measure happiness instead? One nation on Earth has made a name for itself by promoting a measure of National Happiness ahead of a measure of National Wealth and that is the small kingdom of Bhutan. For many years, Bhutan has asked its population through detailed questionnaires about nine aspects of its life (psychological wellbeing, health, education, time use, cultural diversity and resilience, good governance, community vitality, ecological diversity and resilience, and living standards) and combined the responses, treating all nine subjects equally, into a measure of societal wellbeing.

There are questions about Bhutan's motives for doing this, the accuracy of their methods, and various other gripes and groans, but at least they've had a go. Bhutan does behave as though it is trying to make its population happier – and not necessarily through making them richer in monetary terms but across these nine aspects of life.

There is a Happy Planet Index (in fact there are a few similar indices) which attempts to measure sustainable wellbeing by looking at how countries deliver long, happy lives using the Earth's limited environmental resources. Is that happiness? Well, it's a different take on things than simply looking at money. Bhutan comes in the top half of that list at around number 56, with the UK at number 14. Most other European countries are below the UK on this measure, but the list is topped, consistently topped, by Costa Rica. Let's all move to Costa Rica then!

The HPI puts Costa Rica above the UK because despite a slightly (only very slightly) shorter life expectancy in Costa Rica, and a lower (not very much lower) overall index of wellbeing, the ecological footprint of Costa Ricans is lower than that of us, on average, in the UK. I was surprised to find that the French and Germans live a little longer than we Brits, but not only is their wellbeing lower but their ecological footprint is higher. One can argue about the details, but such comparisons are stimulating, I find.

Apart from money, another thing that we certainly count is ourselves. In the UK, the first census was in 1801 and we've been doing them, usually at 10-year intervals, ever since. Only world wars have nudged the timetable slightly. A decennial census is maybe just a habit, but no-one seriously suggests that we should ditch it – people are important, and so the numbers of them are important. The categories of information collected by the census have broadened over time but the information collected in 1841 was pretty similar to that collected ever since, with a few embellishments and a general increase in precision over time. The basis of the UK census is to find out who people are, how old, of which gender, where they live, whether they are married and how they spend their time (in or out of education and paid employment). It seems that if you know that about a nation's people you know a lot, and it's useful.

And it's interesting. In 1901, the population of the current UK area (remember that in 1901 what is now the Republic of Ireland was included in the UK census) was 38,237,000 souls. That suggests that we think we know the population to the nearest thousand and to five significant figures. In 2011 the population was 63,182,000. That growth has been brought about by birth rates being above death rates throughout the whole period (even though birth rates have fallen in absolute and per capita terms, but death rates have fallen faster) but also because the last two decades of data show very high net immigration levels compared with all other decades since 1900 – in fact the first three decades of the twentieth century were decades of net emigration, presumably largely to the New World and to the 'colonies'.

And so to wildlife. Where does wildlife fit in? Well, we don't have an established census of wildlife in the UK in which all species are counted, even roughly. And we don't have a UK National Wildlife Index, and so we certainly don't get regular high-profile updates on its value from government or in the media. There used to be something that approached that in the past. The last Labour government introduced a set of Sustainable Development Indicators in 2001, which covered aspects of the state of the environment including wildlife, the pressures on it, and the responses being taken to adjust things. These indicators were reported on by government ministers every year, with some accompanying media coverage. After the coalition government came to power in 2010 these measures were downgraded, and although some of the measurements are still collected and published, they aren't much talked about, so we can assume that they don't really matter to anyone in power.

There are difficulties in measuring wildlife. One of the main ones is the number of species involved. Over 70,000 species of animals, plants, fungi and single-celled organisms are found in the UK, and my estimate is that only a few hundred of them are monitored every year across the whole country with any value. Most of those species are birds. Would you be happy with using an index of less than 1% of UK species, many of them birds, to represent the state of UK wildlife?

So maybe it's not surprising that we don't see a wildlife index prominently reported and earnestly discussed every day, annually or even decennially. It's a big job and the data aren't up to it. Of course, that's our choice. If you started afresh and considered monitoring financial measures you would probably blanch at the scale and complexity of the task – I'm not sure that measuring wildlife trends is really any bigger a job. But we haven't done it and that speaks volumes about how important we have thought wildlife to be, as a society. Our species hasn't acted as though it needs to know whether other species in the UK are increasing or decreasing in population levels, or why – and that is a strong indication that, as a society, we haven't cared very much.

Wildlife in space and time

An assessment of the UK's wildlife usually involves people going somewhere, at some time, and counting some wildlife. If they make a note of the places and times and the wildlife, and combine their observations with those of others, then we get useful data.

The two bird recording schemes I've already mentioned as a participant, the Big Garden Birdwatch (BGBW) and the Breeding Bird Survey (BBS), have these characteristics. They both require the recorders to say where they are, they both require counts (with rules about what is countable), and they both require the counts to be done at specific times so that everyone is counting over the same period. They both trust the observers to know what they are looking at (or listening to), although there are some safeguards in place in both schemes to weed out misidentifications and/or typing errors. They both ask for some fairly simple information on habitat as well as species, and they both allow, but do not require, observers to record a few non-bird taxa while they are at it. There are two big differences between them. The first is in participation numbers: BGBW upwards of a million people each year and BBS about 3,000 or so. The second is the latitude given to observers over location: in BGBW it's mostly gardens, and every garden is welcome whether it be tiny or huge, whereas the BBS involves monitoring a sample of randomly chosen 1 km National Grid squares, which may contain a mixture of land uses, and participants must choose from the list of available squares.

There is no doubt that the BBS is the more valuable, scientifically, of the two. The BGBW hasn't told us anything about changes in numbers of breeding Lapwings or Pied Flycatchers or Skylarks, and it never will because it's carried out in winter and in gardens (the clue is in the name), but, unlike some of my colleagues, I wouldn't be too sniffy about BGBW – because analysis has shown that for those species where there is overlap (resident widespread birds that occur in gardens) then there are similarities in the results for the two schemes.

There is another important similarity between these two schemes – they have both been running, with some changes, for decades. Just as the best time to plant a tree is 20 years ago, and the second-best time is now, the same applies to monitoring wildlife. That's because, like trees, censuses take time to grow and blossom, and so it's best to get on with it – but it's also because we are actually much more interested in changes than in absolute numbers. One of our problems in assessing what is happening to our wildlife is that very few people or organisations put the work in 20 years ago, or 50 years ago, that would be worth its weight in informational gold today. Any conservationist gifted a time machine should consider going back a few decades and setting up plant, insect and marine monitoring schemes that would serve us well these days. The pay-off from monitoring mostly comes years down the line, and so to start such a scheme is a vote of faith in the future that it is worth doing at all – because it depends on our expectation that there are changes afoot and we want to know about them, and perhaps do something about them. If you were, let's just imagine, the government, you might be rather glad that there aren't too many monitoring schemes around – because if there were it would be far easier for you to be held to account.

In order to know how well wildlife is doing in the UK we don't need to know the address of every creature in the country, but we often would like to know the relevant information for the UK, its four constituent nations, regional areas such as counties and often the site where the numbers occur. For example, we might want to know how many Bitterns there are in the UK, in England, in Suffolk and, for different reasons in each case, at the RSPB's Minsmere nature reserve. Being able to say where things are is not only useful for visiting them, but also for understanding why their numbers are as they are – and crucial for implementing conservation measures for them. And often we don't need to know the absolute numbers of the things we count, we want to have measures of change.

What should we count? The obvious and generally correct thing to count, in my view, and I hope it has shone through the words that you have already read here, is the abundance of each individual species. Wildlife presents itself to us as recognisable things called species. When I do my BBS survey, I record every bird I see or hear and recognise, but I don't note them all down simply as 'birds', I note them down as Skylarks, Blue Tits and Woodpigeons, and that's a simple task for me. You could tell the difference between a Mute Swan and a Robin (and so can I) and it makes no sense to lump them together as birds, or lump them with the occasional Rabbit, Frog or Brown Hare as vertebrates: it's interesting to know whether Mute Swans are increasing or decreasing and not just to know what vertebrates are doing as a mass.

I'd be competent, and fairly highly competent, at doing a BBS survey anywhere in the UK, although if you put me in an oakwood in Wales I'd need to refresh my memory of the songs of Redstart and Pied Flycatchers (but I'd be fine with Tree Pipit and the blissful song of the Wood Warbler). But tell me to identify plants or most insects, even on my home patch, and I'd be no use to you.

Could a more competent all-round naturalist simply count the plants and insects as they were counting the birds on a BBS visit? Not really, and they'd struggle to count the mosses, spiders and earthworms while they were at it. If that were where the bar was set then you'd have to count me out and, quite frankly, I don't think you'd be left with many recorders.

If we want to know what is happening to all 70,000+ UK species then we need a lot of expertise to do it, and probably a lot of time and a lot of money. There is no day set aside every decade to do the wildlife census, as there is to do the human census. That is understandable, but it's also somewhat disappointing. But if we could do it, then we'd certainly want a measure of many species – not necessarily all 70,000 but lots of them, and not just lots of birds, butterflies and a few flowering plants.

Is there an easier way to arrive at a health-check of the UK's wildlife? Well, there is a complementary way, although I don't think it could ever replace or be as useful as assessing the status of species, and that is to assess the extent of habitats. But what is a habitat? That's a good question, and one which is far more difficult to answer than 'what is a species?' (although that too is a somewhat fraught question). Habitats tend to be named after the predominant plant types that are their very foundation: woodlands certainly have woody trees in them, heathlands usually have heather-like shrubs in them, and grasslands tend to have lots of grass. These are incredibly useful terms, as they conjure up for us, immediately, a vague impression of what the area in question is like – but the definitions get pretty fuzzy around the edges, especially if too many botanists are involved. And we do have a problem that the UK has rather little natural habitat covering its land surface in any case, so we then have to list semi-natural habitats, which include farmland – which in turn seems less like a habitat, or even a bunch of habitats, than a land use.

We refer to habitats a lot of the time, but it is incredibly difficult to discuss wildlife in the UK without focusing on species. If I'm told about an area of chalk grassland I will almost certainly ask which species live there, maybe which species of butterfly, and I'll judge the chalk grassland partly by the species it accommodates and also by whether they are at high or low densities and whether they are increasing or decreasing in numbers. Habitat extent is good to know, but habitat quality is got at by knowing how species are faring in that habitat. The fate of species is fundamental to our assessment of the overall state of wildlife in the UK. Essentially, it is the fate of species that keeps us honest. If we protect the habitat, or draw lines around areas and call them wildlife reserves or protected landscapes, and if the species in those places plummet in numbers, then we aren't doing well enough, however much habitat there is on the map.

Wildlife monitoring schemes are useful to have, but challenging to establish, and it is noticeable that many of the existing ones have been set up by non-governmental organisations (NGOs) and not by government (although government agencies very often help to fund them, they do not run them). In contrast, the human census is run by the state, and is compulsory, it's not set up for the public good by a charity. We really do show what we care about through what we measure, monitor and report.

The gold standard?

When I write about monitoring schemes that tell us about the state of wildlife, I tend to write about bird monitoring schemes. That's because they are what I am familiar with and they are, to a large extent, at the top of the tree in terms of power to inform. I often hear people saying things along the lines of 'bird people are so lucky to have all those data', and to some extent they are right, but mostly they are wrong. And luck had nothing to do with it.

Birds have some inherent advantages if you want to monitor their numbers, among which are: there are lots of birdwatchers; bird identification is usually quite easy; birds draw attention to themselves by singing and flying around; and in any one place there is a manageable number both of species and of individuals to record. But they have some inherent disadvantages too: they occur everywhere, so coverage has to extend from central London to the tops of high mountains; they fly away from the observer; some are silent or unobtrusive; they sometimes occur in staggeringly large numbers; there are different species present at different times of year. You could add to both lists.

The BBS scheme, for which I carry out surveys each year, was established in 1985, but that wasn't where recording of common birds in the UK in the breeding season started, because the BBS was specifically designed as a replacement for an existing scheme, the Common Birds Census (CBC), which had existed since the mid-1960s. The CBC was conceived at a time of growing environmental awareness, of concerns over pesticide impacts on farmland birds, and after the 1962/63 winter, which was the harshest in terms of prolonged freezing weather and long-lying snow since 1740. Resident bird populations plummeted, with perhaps as many as half of the UK's resident birds dying that winter. The next spring's dawn chorus was much quieter, and you didn't really need a monitoring scheme to tell you that Wrens, House Sparrows and Blackbirds were much rarer than normal.

The CBC was a good scheme, and if we now had anything really comparable in scale and design for other taxa, 60 years after it was set up, then we'd be very pleased. But although the CBC was the gold standard in the 1960s, by the 1990s it looked made of a baser metal – such is progress. The two main problems with the scheme were the small sample of sites covered – just a few hundred – and the non-random nature of the site selection, with most sites consisting of farmland and woodland, mostly in England, and including quite a lot of wildlife reserves in the sample.

The CBC was also time-consuming, involving around half a dozen visits to each site each year, detailed transcription of the sightings onto maps, and then complex analysis of the maps by experts to assess population numbers. This limited the attraction of the scheme to the volunteers who collected the data – it was a lot of unpaid work – and increased the costs of analysis for funders. But also, volunteers could choose their own sites and, not surprisingly, tended to choose sites that were good for birds and easy to get to, so there were doubts about how well the data would represent what was happening generally in the countryside. Those doubts may have been slightly overblown because it proved possible to knit together the results of the BBS and CBC, as if by magic, by having some overlap years when the two schemes ran alongside each other. The BBS, a simpler scheme for volunteers to carry out and for the organisers to manage, now covers scores of species and produces viable indices of population change for a decent number of them, not only for the four UK nations but also for many regions of England where the density of volunteer observers is high.

The BBS is a breeding-season survey and is well adapted to provide information on common and widespread species. Other bird monitoring schemes, some annual, some not, have been designed to assess wintering bird numbers and trends (particularly of waders and wildfowl), birds of prey, rare breeding species and breeding seabirds. All in all, we have very good information on trends in bird numbers. Those various schemes, taken together, represent the gold standard.

The true test of the adequacy of monitoring schemes is whether they can produce graphs of a measure of abundance over time. The more dots the better, the further back in time they go the better, and the more that you can believe in the veracity of the individual dots the better. We have a lot of good graphs for birds, and rather few for the rest of our fauna and flora.

Ornithology shouldn't be too bashful about the quality of the bird information, and botanists and entomologists might look hard at whether they have done enough to get into a better position in the last decades – I think they haven't.

But let's put all that aside and think for a moment of an imaginary world where instead of the existing levels of information (good for birds, hopeless for the marine environment, not great for most plants and invertebrates, pretty patchy for other vertebrates) we could choose one of those groups to displace birds as the group of organisms about which we would most like really good information. Which would we choose? Difficult, isn't it? I could make out a case for any of them really.

I'd really like it if we had much better information on loads of other taxa. In an imaginary world, I'd be prepared to trade away some bird information in order to get more information on plants and in-vertebrates for sure, but that Faustian deal is not on offer.

UK extinctions

There are extinctions and extinctions. Global extinctions are pretty big events, the end of a species on this planet, but national or local extinctions are often the loss of a species on one side of a slightly arbitrary line whilst things are going much better on the other side of the line. You could even say that species go extinct in my garden several times a day, but because they quickly recolonise I don't worry too much about it. When it comes to extinction, one needs a sense of perspective.

The Aurochs that were ancestors of domestic cattle once roamed Britain, but they've not roamed anywhere on the planet for about 3,500 years now – they are extinct. The Large Copper butterfly must have been quite common in the Fens of East Anglia and elsewhere in southern England in the distant past but was last seen in the UK as a native species in the late nineteenth century. Because it remains widespread and quite common in various parts of Europe we should regard this as a UK extinction (I prefer the term extirpation) to distinguish it very clearly from a species that has disappeared off the whole planet. And the Coal Tit that was on my bird feeder a moment ago but has flown over the fence will be back soon, so we don't need to call that anything at all.

The list of species formerly occurring in the UK but which now don't occur on planet Earth is very large if you take the longest possible view. Beneath my feet when I collect the milk from the doorstep are Jurassic rocks laid down 160 million years ago, when ammonites and ichthyosaurs swam the oceans, but they are long gone from the planet. As are the much more recent hyena, rhinoceros and mammoth species of a mere 200,000 years ago. More recently, the Great Auk was found in the UK into the nineteenth century but had disappeared here as a breeding species decades before the last two individuals were killed in Iceland in 1844. However, I can't find an example of a species whose last living member, pre-extinction, was found in the UK. The UK has not been the final refuge for any species on its way to global extinction, and nor have any of the endemic UK species, those found nowhere else in the world, gone extinct as far as we know.

The UK has few endemic species – although it does depend on where you draw the species boundary. This is largely because the impact of several ice ages, the last about 10,000 years ago, was to wipe clean the UK fauna and flora, with the result that most of the species we have are ones which have recolonised from continental Europe since then, so they are unlikely to have had the time and the conditions to evolve into UK endemics. On the other hand, the existence of the English Channel has meant that there have been a few thousand years during which isolated UK wildlife could evolve in slightly different directions from their parent species. The net impact of these two factors is that there are few endemic species but many endemic subspecies – which perhaps, given more time, will evolve into UK species.

There are fewer than 100 UK endemic species and the majority of those claimed, mostly plants and fish, are of disputed taxonomic uniqueness. They comprise a clutch of whitebeams, sea-lavenders and charrs, some of which have very small geographic ranges and populations. Less controversial examples of endemic species include the Lundy Cabbage, which is only found on that island at the mouth of the Bristol Channel, and if it were ever to disappear then so too, presumably would the Lundy Cabbage Beetle and Lundy Cabbage Weevil, which both depend on it. If you are looking for an endemic UK species in which you can delight then I reckon that the Scottish Primrose is the best example. It is found along the coast of Caithness and Sutherland, and on Orkney, and is a pretty flower that is easily identified – I'd be very sorry if we were to let it slip away from the planet and it is entirely up to us, particularly Scotland, whether or not it does.

But unlike the Lundy Cabbage and its beetle and weevil, restricted to one island, and hanging on there at present, the Brown Bear which last roamed these parts a good 1,000 years ago is gone but lives on in other parts of Europe – where indeed they are increasing in numbers and being translocated back to places from which they were extirpated more recently than their UK demise. We could attempt to bring them back – it wouldn't be high on my list of wants, though maybe you are keener on the idea than I am. But once you get into the list of extirpations (gone from here, still extant elsewhere) then there are loads of species on the list, and some of them already provoke fierce debate over whether they should soon be, or should never be, reintroduced.

The four top mammals for possible reintroduction are Brown Bear, Grey Wolf, European Lynx and Beaver. For me, Beavers are no-brainers – we'd be better off with them than we are without them and we shouldn't let prejudice and nervousness get in the way of restoring them to our countryside in a big way. Brown Bear I can very happily live without. Grey Wolf and Lynx I would love to have back, even if I never saw a single one myself, but I think they would do good to our ecology – they'd reduce the numbers of deer and probably of several medium-sized predators (including domestic cats) and they would be little cause for concern to humans, either in terms of physical danger or because of their impacts on livestock. I'd be very happy to be guided by experts, but I have a feeling that the parts of the UK where both species would do the most biological good are exactly the places where they would stand the least chance of becoming established because of human disturbance and the toll that cars would take of their numbers.

Extirpated birds ripe for reintroduction? Well, most could get here on their own, perhaps, but from the distant past the Dalmatian Pelican was once found in the Fens, the Somerset Levels and similar haunts and is now a globally threatened species for which an increase in world range to former localities could only be a good thing. Would it work?

Species reintroductions are one of the mildest forms of rewilding – they simply represent putting species back where they have been extirpated by us in the past. And with the existence of a marine barrier to continental Europe there is little prospect of many species making it back on their own. Among the successful reintroductions of recent decades the White-tailed Eagle and the Beaver are the big two, and both of them currently have very restricted ranges and are still the subject of some controversy.

Let's play a thought game. If you could bring one extirpated species back to this country, what would it be, and if you had to sacrifice another native UK species from the same group (a bird for a bird, a mammal for a mammal, and so on), what would you choose? A few years ago, Beaver might have been a common choice of species to bring back, but maybe today we think it is already well on the way to being back – or do we? Let's just use the Beaver as an example. Which native mammal would you sacrifice to get the Beavers back? There are quite a few non-native mammals I'd be happy to see the back of, but Grey Squirrel, Muntjac Deer and Brown Rat aren't in play in this game. I think I'd let the Greater Mouse-eared Bat go. Now, I like Greater Mouse-eared Bats. I've seen them abroad and I like them very much, but they are at the edge of their range in the UK (although, to be fair, geography makes that true of many species). But maybe that's an easy choice. Would I let the Wildcat go in return for the safe establishment of Beavers in the land (and the water)? After all, the Wildcat seems pretty close to extinction anyway. I think I would actually. Please don't hate me for it.

Maybe we pay slightly too much attention to extirpations and even to extinctions. Most national extinctions are pretty trivial events, sad events and not events to be wished for, but neither earth-shattering nor nation-shattering in impact. The Skylark is never going to go extinct in the UK (probably) but we have lost millions of them from our countryside in my lifetime. We've lost scores of pairs of Red-backed Shrikes in that period and they are, more or less, extinct as breeding birds in this country. Given the choice, I'd rather have the Skylarks back than the Red-backed Shrikes. I'd like both, but I'd put the larks ahead of the shrikes. Why? It's largely because Skylarks seem to me more of what makes the British countryside the British countryside than do Red-backed Shrikes, so I guess that means it comes down to a cultural choice. Of course, I get more birds for my money if I opt for Skylarks, but I also get a more important restoration of what we have lost, it seems to me. I'd rather see Skylarks go back to 100% from around 50% now, than see Red-backed Shrikes go to 100% from their current level of 0%. Although this one is hypothetical, wildlife conservation is all about difficult choices.

Non-natives

As well as the species which used to live in the UK but no longer do, there are species that were never native to these parts but now have self-sustaining populations. These include some that were deliberately introduced in the past by people thinking that they would be assets to our land (e.g. Rabbit and Brown Hare), ones which have escaped from captivity in zoos or gardens (e.g. Muntjac Deer and Ruddy Duck), and those which have arrived accidentally through trade (e.g. Brown Rat, Mitten Crab and New Zealand Flatworm).

Non-native species are often, but not always, conservation or economic problems. A global perspective would deliver lots of examples of both – global extinctions of island species caused by the introduction of diseases and predators to ecologies that had developed without them, and species that have done so well for themselves that they have harmed the economic wellbeing of industries, particularly agriculture.

But it's far from the case that all non-native species that are tipped into the UK countryside, rivers or seas become problems. For a start, most of them don't survive and don't increase in numbers. The numbers of escaped budgerigars each year must be considerable, but all those escapes have not resulted in a feral population, probably because the UK only tangentially resembles the great Australian outback from which the bird hails. And many that do establish themselves are pretty benign additions to our wildlife (at least, as far as we've been able to tell so far). There is a rough rule of thumb that one in ten introduced species becomes naturalised and lives in the wild and one in ten of those becomes a significant economic or conservation problem.

But when they are established, and are causing problems, it is mighty difficult to get rid of non-natives. There are two examples of deliberate planned and successful removal of non-native species from the UK in recent years: Coypu, a mammal from South America, and Ruddy Duck, a bird from North America.

Coypus are large semi-aquatic rodents whose native range is eastern South America. They are large, 4–9 kg (similar to a Red Fox or Otter, and at their largest not much smaller than a Badger), and therefore are fairly immune from predation in the UK countryside. Ironically, their numbers were reduced by overhunting in their native range because the species is prized for its thick fur, and so Coypu farms were established in many parts of the world from which, as almost always seems to be the case, animals escaped into the wild, in the UK in the 1930s. Between the 1940s and 1960s there were several attempts at eradication, because Coypu cause damage to waterways, and then the 1962/63 winter clobbered the population, which was restricted to East Anglia. But it was not until a well-resourced, well-planned and long-term eradication scheme was implemented in the 1980s that the last Coypu swam in a British river.

Ruddy Ducks appear to have been eradicated from the UK during the 2010s. This attractive American duck was released accidentally (there are rumours that it wasn't so accidental) from the Wildfowl & Wetlands Trust Slimbridge Centre, in the days of its founder, the great environmentalist Sir Peter Scott, in the 1950s. Feral populations were established in the West Country and West Midlands and the species became a rapidly increasing sight as a breeding species and winter visitor to wetlands across the country. The trouble was that Ruddy Ducks left the UK, particularly in cold winters, and came into contact with the only related European species, the White-headed Duck, which was endangered in any case, and all the more so when it was found that, somewhat surprisingly, hybrids between Ruddy and White-headed Ducks were themselves fertile. The strong prospect of genetic swamping of the European species by the rapidly increasing North American import led to a controversial UK eradication programme, which appears to have been successful.

Two expensive national eradications and scores of potentially damaging non-native species leads one very quickly to conclude that prevention is better, and much cheaper, than cure, but global trade, timber imports, garden plants and many other sources are very difficult to control and so laws have been tightened – but it feels like a pretty hopeless and daunting task to eliminate the many new non-natives, especially plants and diseases, that reach our shores. Can we ever, realistically, expect to see the last of Signal Crayfish, Japanese Knotweed or New Zealand Pygmyweed?

Doing anything about non-native species is difficult, and demonising the species in question is not the way forward. There is nothing wrong with Grey Squirrels, it's not their fault or their choice that they are here causing problems. But even talking about the problem is fraught, given that it edges towards a type of racist language used of people. There is a very clear difference in that here we are talking about non-native species, not fellow humans of our own species. Across the world non-native species cause problems, not all the species but some of them, and not just little problems but some very big economic and conservation problems. You must make up your own mind what you think about human movements around the globe, but transfers of wildlife to areas far outside their native range are definitely to be avoided.

Woodland

Most of the UK wants to be a forest – woodland is the climax vegetation type for all but the coldest, highest or wettest sites in the country, and even in our uplands the natural tree line would reach up to the tops of all but the highest mountains. The 70% of our land cover that is farmed and the 8% that is built up would all go back to woodland if we upped sticks and left wildlife to get on with it for a couple of hundred years. And yet the UK is one of the least forested countries in Europe – partly because we don't have many very mountainous areas that people have left alone for centuries, but mainly because we do have a lot of land that is good for farming once you cut the trees down. And we have quite a lot of urban areas and the road and rail infrastructure that go with them. That's why the UK brought down the average woodland cover for the EU quite a lot – our 13% cover was well below the 41% EU average (in 2018).

Woodland cover has more or less doubled in the last century, and there are government targets to increase it much further as a climate change measure over the next decades. A large part of the doubling of woodland cover has been achieved by the setting up, just over a century ago, of the Forestry Commission, a government forestry service (now devolved in different ways across the four UK nations) which bought land and planted trees. Many of those new woodlands, often in upland areas, are plantations of non-native trees that are commercially attractive because they are fast-growing. But because they aren't native even to this continent, and they have been planted in dense tree-farms, just like many an arable crop, they leave little room for wildlife compared with native non-commercial woodland. Plantations of Sitka Spruce, which hails from west-coast North America, are far from wildlife deserts but the species that live in them, a bit like those that live in the middle of intensively farmed wheat fields, are generally a small subsection of generalist species.

The best woodlands for wildlife, generally speaking, are those that have been around for a long time and whose flora and fauna have matured over time. These are the woods in which the UK's 'real' woodland wildlife has been allowed, and in some cases encouraged, to find refuge and thrive. The specialised woodland plants and bugs are mostly found in old woodland sites, even though they must have been widespread and common species centuries ago.

How much of our woodland is old? It depends what you mean by old. There are ancient woodland inventories for all parts of the UK, but ancient woodland is defined as that existing since 1600 in England, Wales and Northern Ireland whilst 1750 is the benchmark year in Scotland – the reigns of Elizabeth I and George II respectively.

Given that something like 90% of our land area wants to be a forest, how much of that original land area has persisted as forest throughout, and meets the 1600 or 1750 qualifying criterion? The answer is about 2.5%. And during the last century, while we have doubled the area of woodland cover, we have halved the area of ancient woodland, which in retrospect doesn't seem very clever. And things get worse, because even that 2.5% coverage includes areas with no old trees at all and practically no native trees, woodlands that have been felled at some stage and replanted with non-native species.

To add some further context to this, the woodland cover in the late eleventh century, as recorded in the Domesday Book (England only, and most though not quite all of England) was around 15% – fairly similar to the current level.

Ancient semi-natural forest in the UK is the equivalent of old-growth forest in North America – they have a better name for it than we do. If you look at a map of our ancient forests then you see the remaining ancient woodlands as islands in a sea of urban development and farmland. The natural vegetation of this country is reduced to remnants, small remnants, isolated and beleaguered remnants – but how is wildlife faring in our ancient and modern forests overall? Not so well.

It's worth starting with the plants, particularly as Plantlife produced a stimulating report, *Forestry Recommissioned*, less than a decade ago which identified three main factors affecting woodland plants: lack of management, nitrogen deposition and browsing by deer. Given that plants are at the base of most food webs, what affects plants clearly will affect everything else, so it's worth taking note. We have chopped down most of our existing woodland, some of which we've replanted, and people have been harvesting timber for building, firewood, fencing and other uses throughout human existence – so it isn't surprising that management of woodland is important to what wildlife thrives there. Our woodland wildlife has spent hundreds of years coping with, and adapting to, how we use woodlands, and changes in management and lack of management determine woodland flora and fauna. The report carries a striking statistic to illustrate change: in 1947 English broadleaved woodland was half and half – half high forest and half coppice or scrub – but in 2002, 97% of such forest was high forest. That is a stunning change in management intensity and has major implications for how much light reaches parts of the forest and which species of butterfly and other insects will thrive.

Nitrogen deposition is the second factor highlighted by Plantlife – the equivalent of fertiliser falling from the sky. About 95% of our woodland is growing under conditions of excessive nitrogen deposition, in the form of nitrogen oxides and ammonia, coming from sources outside the wood itself. This fertiliser favours plants such as Nettle and brambles – fine in their own way, but they tend to take over the woodland floor and crowd out other plants so that we have homogenised woods with little variety in plant life. Nitrous oxides come mainly from the burning of transport fuels and fall as rain, so areas with high transport levels and/or high rainfall get the brunt of the impacts, and in a crowded rainy country that means almost everywhere except the remote northwest of Scotland, which is protected by the prevailing winds coming from the Atlantic and a low population density, even in the tourist season. Ammonia mostly derives from agriculture, particularly fertilisers both organic and inorganic, and again can come from local sources, such as a large chicken unit's manure, or from distant sources, carried by clouds and deposited in rain.

Lichens, liverworts and mosses are particularly sensitive to nitrogen deposition, and the UK is internationally important for them, largely because of our wet temperate climate. However, large areas will be much poorer in these special plants than they should be because of aerial pollution.

Deer are a problem for woodlands – they limit the establishment of trees and browse away the understorey. We have high densities of deer in the UK because we lack the large predators that would limit their numbers, most notably Grey Wolves, Brown Bears and Lynx, but also because landed gentry in the eighteenth and nineteenth century shovelled out a load of non-native deer species, notably Muntjac, Sika, Chinese Water and increased numbers of Fallow, which have thrived in the absence of predators and any other effective form of control. High densities of deer have changed the ecology of Britain's native woodlands, particularly in the south of England, and much wildlife has suffered as a result. Deer grazing has not just affected plant species, but the large structural changes are the causes of some bird declines too.

How are birds faring in our woodlands overall? Can they shrug off the lack of management, the nitrogen deposition and the impacts of deer? Not entirely. Woodland bird numbers have declined by about 40% in about 50 years. This is certainly due to the three factors identified by Plantlife from a plant perspective, but many of the declining woodland birds – the Spotted Flycatcher is one of them – are long-distance migrants, and so they may be affected by all sorts of factors operating on their wintering grounds far away in Africa.

Global warming affects phenology, the timing of seasonal events such as flowering of plants, growth seasons and emergence times of insects. Spring is getting earlier – on average by about 26 days earlier than it did in the period before 1986 – and lasts about a month longer. This has implications for food chains. Oak trees come into leaf earlier in early, warm springs, and the caterpillars of moths that feed on oak leaves have adapted to emerging early too to continue to cash in on the feast, but those caterpillars are also an important food source for woodland birds, some of which find it easier to shift their nesting seasons to remain in sync than do others. Generally speaking, resident birds are coping OK, short-distance migrants are coping quite well, but long-distance migrants appear to be less able to rush back to the UK weeks earlier – and some of the fastest-declining woodland birds are long-distance migrants.

In summary, the state of the UK's woodlands for wildlife is pretty dire. Woodland once covered a large slice of the land, but we have cut most of it down, and the rest of it is poorly managed, drenched in harmful aerial fertiliser, under assault from native and non-native deer gobbling everything up, and threatened by climate change disrupting woodland food webs. The status of woodland wildlife, and the pressures it faces (not all of which have been described here), is a case study in the struggle for existence in current-day UK.

Farmland

The farmland that has replaced much of our native woodland is of a wide variety of types, from arable land growing crops such as cereals, oilseed rape and beans in the eastern areas, to dairy farms in the wetter west and predominantly beef and sheep regimes on the higher ground, with dwindling areas of mixed farming enterprises in some parts of the country. When we think of the British countryside most of us probably have in mind open fields with scattered areas of woodland – and this seems normal, indeed it is normal and has been for centuries, but it is a normality created by us imposing our land use on the natural ecology of the land.

Our domestic animals have been selectively bred over many years to provide high yields of milk, meat, wool and eggs. Few of our domestic breeds would thrive if released into the countryside to fend for themselves, as they depend on foodstuffs many of whose components are imported from faraway countries and veterinary medicines which allow high densities of animals to be kept, particularly indoors.

Our primary arable crops are intensively bred plants from all over the world which are fertilised, treated with a wide variety of '–cides' (herbicides, fungicides, insecticides) and sprayed with growth regulators in the fields. Even the grass grown for livestock isn't natural grassland but intensively fertilised grass leys of a few species which are grazed in summer but often cut for silage in late summer, with the winter forage stored in huge plastic bags to ferment and be eaten months after it has been harvested.

Very clearly, farmland is not a natural habitat. And the closer one looks at these familiar fields the more it becomes a matter of wonder that much wildlife survives at all in our farmed countryside. But in terms of food production, this is a very successful enterprise. Yields of most farm products are high in the UK from a combination of good soils, favourable climate and highly intensive farming practices. The top four wheat-producing countries in the world (China, India, Russia and the USA) are all much bigger than the UK but all have far lower wheat yields per hectare than we do: 5.4 t/ha, 3.4 t/ha, 2.7 t/ha and 3.3 t/ha respectively, compared with the UK's 7.8 t/ha (in 2018). Wheat is sold on world markets and so, to a large extent, a wheat farmer in Northamptonshire is getting the same price for each tonne of the crop as one in Nebraska. A doubling in yield is therefore pretty significant, even though the Northamptonshire and Nebraska farmers will have different costs of land, labour and machinery – but we can see that it would be a strange world in which the farmers of a country which can grow some of the highest yields of wheat in the world were to decide to grow rice instead. Our farmers are not mucking about when it comes to wheat growing, and estimates of attainable yield, under current circumstances of technology given prevailing soil and climate conditions, suggest that they are pretty much top of the league in squeezing as much wheat from their fields as anyone else in the world – they are intensive wheat growers, and highly efficient ones in agricultural terms.

And moreover, if we were to look at most other agricultural statistics, UK farmers are very good at producing just about everything: cereals, oilseeds, vegetables, milk, eggs and meat. Farmers will tell you how efficient they are when they really mean productive because we are talking here about outputs (yields) and largely ignoring inputs (fertilisers, medicines, pesticides, labour, machinery). On the other hand, if you ever find yourself in conversation with a farmer about how rich they are, given that they are one of the most 'efficient' and productive farmers in the world, then you get a different story about how tough farming is and how they aren't making money out of it because life is so hard. Both can be true.

What can't be denied is that UK farmers have been some of the most efficient in Europe at driving wildlife from their, our, land, and that UK farmland is much less productive for wildlife than it was 50 years ago. Is this just bad luck, or are the farming techniques that put us high in the lists for yields the very same ones that put us low in the lists for wildlife production? It seems that they are.

Twenty-five years ago, one heard a lot about declining farmland birds, but it seems that, despite little recovery in numbers, this is old news and that a dreadfully birdless countryside is the new normal. Of the 70 bird species on the UK Red List because of their poor conservation status, a good third of them are species of upland or lowland farmland – but perhaps that's not surprising given the extent of farmed land in the UK. Perhaps more telling is that the number of farmland species on the list keeps growing, species are often added and rarely removed, and that many species, for example Turtle Dove and Corn Bunting, look more likely to exit the list through UK extirpation than through a recovery to be downgraded to the Amber List.

There was great intensification of UK agriculture in the 1970s, and this led to declines of farmland birds during that period and through the 1980s and 1990s. Not only are there innumerable studies that document the role of agricultural intensification in causing bird declines, species by species, but comparative studies that demonstrate, across the European continent, that the countries with the most intensive agriculture have lost their bird populations the most rapidly. We are now at a stage where countries in southern and eastern Europe, which used to support rich farmland bird populations, are catching up with the UK in ridding their fields of song.

The well-documented bird declines were probably preceded and accompanied by large declines in invertebrates and plant species, quite probably of an even larger magnitude than the avian declines. The 70% of our land area which is farmed, most of our country's land area, is where wildlife has drained away from our lives. And yet agriculture is massively subsidised from the public purse and has grants to protect wildlife poured into it in huge quantities. If we are looking for an example of public policy failing to deliver a balanced outcome we need look no further than agriculture.

The marine environment

Considering that the Earth is a blue planet, with two-thirds of its surface covered with salty water, we have remarkably little information on the changing abundance of the myriad creatures which inhabit that realm, and that is the case with the marine environment around the shores of the UK.

We know quite a lot about the changing status of seabirds but that is mainly because they come to land to nest, and their status and changes in numbers have been driven partly by what happens to them in those brief partly landlocked months – so they form a complex and slightly frosted window into what is happening in the marine environment. The top line on seabirds in the UK is that the UK has very important populations of many seabird species, with our Gannets, Manx Shearwaters and Great Skuas numbering over half of their respective world populations and impressively large colonies of many other seabird species. These seabirds are summer visitors to our shores, and indeed our seas, with many of them wintering in the southern hemisphere. But these impressive seabird populations tell us that the seas which can be reached from our shores are rich in food as well as that our cliffs and islands provide relatively safe locations for raising young, safe from predators and nowadays largely from the harvesting activities of our own species.

Apart from seabirds, the changes in marine wildlife that we have measured quite well are largely limited to changes in the catches of commercially important fish (and previously cetaceans) and in more recent decades the abundance of tiny marine creatures which form the base of many marine food chains.

The overall picture which arises from a review of marine fisheries in UK waters is that there aren't as many fish in the sea as there once were. Fish stock after fish stock has collapsed, due to overfishing. Fishermen (as they are mostly men), across the globe, argue until they are blue in the face that they aren't overfishing and that there are plenty of fish out there somewhere, and if there aren't then it is someone else's overfishing that is the problem, but the truth is that we have overexploited the stocks to the point of commercial incompetence and economic disaster.

Landings of commercial fish cannot really be taken as accurate measures of the numbers of fish in the sea because fishing effort varies so much from year to year. Fishing effort is determined by regulations by governments, the weather, the technology available to travel across the sea surface, find the fish and haul them out of the water, and the relative abundance of other fish species. For all those reasons, and more, the landings of a fishing fleet are far from the standardised observations of visits to BBS squares. Fishing fleets go where they want, when they want (within the regulations) to find and remove as many fish as possible (and as allowed) for commercial gain. And there is no army of volunteers taking to the seas to collect standardised observations of a range of fish species as background to the commercial catches.

But there are big advantages too to looking at landings of fish – the fish were landed, and sold and eaten. These are not paper records of what someone claims to have seen, they are tonnes of fish flesh landed at the quay, and we can expect that the species identification and tonnages are pretty accurate, since money is involved. We can also assume that fishermen are trying hard to catch lots of fish – this is not, for them, an academic exercise. And there are records of landings of fish that go back a long way.

In 1889, the UK fishing fleet landed twice as much fish (by weight) as it does today. I find that an amazing figure, particularly bearing in mind that the fleet was largely restricted to sail as a means of propulsion and the boats fished much closer to shore than they do these days. In contrast, wheat yields per unit area in the UK have quadrupled over the same period. By 1937, technological advances (shipboard freezers and engines both allowing longer fishing trips to more distant seas) meant that British boats were landing more than eight times the weight of fish as in the late nineteenth century and 17 times as much as is landed now.

Over the last century our ability to find, catch and land fish has increased enormously and yet landings have fallen. Most of this change is attributable to overfishing. Whereas, from a commercial point of view, our Victorian forebears were underfishing (in other words, there were more fish that could be caught sustainably), in our lifetimes we have overfished. In addition, and not helping fish stocks, the catching methods have been too damaging to the marine environment for catches to be maintained. Furthermore, but of lesser importance, we have polluted and disturbed the marine environment, and there are changes in that environment that are attributable to warming seas caused by climate change too.

Similar stories of overfishing occur across the global oceans, with very few exceptions once technology is brought to bear on fisheries. Our treatment of the seas, but most particularly our treatment of fish populations that are of immense value to us if managed sustainably, makes depressing reading. If we can't get this right, what hope is there for us to get more complex interactions with wildlife right?

The marine environment has one monitoring scheme for which we would give our eye teeth to have an equivalent on land. The Continuous Plankton Recorder (CPR) is a bit of kit that can be towed behind a boat and extracts plankton, tiny floating animals and plants, from the sea. The first prototype was used by Sir Alistair Hardy in the Antarctic to sample krill in the mid-1920s; it was then further tested in the 1930s and deployed regularly behind ferries, in the North Sea and elsewhere, from 1946. The CPR scores very highly on standardisation and longevity. And the data have been routinely analysed and published.

The plankton consists of phytoplankton, single-celled algae that produce half of the world's oxygen and fix 100 million tonnes of CO_2 every day, and zooplankton, tiny animals which include the larval stages and eggs of larger organisms such as jellyfish and barnacles. These creatures float in the water column and are carried around on ocean currents.

Sixty years of CPR data have shown dramatic changes including northward shifts in plankton abundance caused by warming of the seas, and changes in the timing of cycles such that the seasons occur 4–5 weeks earlier than previously, in a similar way to the changes in timing of egg laying by birds, flight times of insects and flowering of plants on land. These changes in currents and temperatures are the main changes in the marine environment out at sea, with pollution and nutrient levels being largely unchanged, but they represent fundamental changes to the marine environment, because plankton are the food for many fish and cetaceans and form the basis of almost all marine food chains.

I always think of the marine environment as being the poor relative with regard to data compared with terrestrial environments, and we have especially poor information on the fate of benthic (bottom-living) marine species – but this type of information on marine plankton is so far ahead of what is available for aerial plankton, the insects that used to splatter on our car windscreens, that I wish Alistair Hardy had been an entomologist rather than a marine biologist.

Where we sit in the global wildlife crisis

There's nothing like a good crisis to make life more interesting and make us all feel as though we are part of important events. A crisis is an event or time when things are going wrong, or maybe about to go wrong in a pretty serious and damaging way for something or someone. Crises are times when action, the right action, must be taken to steer away from the worst outcomes, and they are when leaders either come into their own or fail.

When I was an undergraduate, I sometimes had essay crises when I couldn't go to the pub that evening because I had one or more essays that had to be finished, sharpish. I have lived through many crises of English cricket, and we are never far away from crises in education, social care, health and employment. My parents lived through the crisis of the Second World War and I am living through the crisis of a global pandemic and another of climate change. The purest of these crises, in their own very different ways, were the university essay crises, the approach and onset of a world war and the initial stages of a global pandemic. All of these had the characteristics that action needed to be taken over a short period of time to reduce the chance of disaster. I think we could call these Pure Crises. The other matters are very important, and we can call the worst-case scenarios disasters (each in its own way), but they are largely of our own making through a consistent and oft-repeated failure to act when the evidence of oncoming worsening of the problem is staring us in our faces. For many of the things that we like to call crises there is no critical (note the word) moment, but a series of moments when we fail to act because we could act tomorrow... but every tomorrow has another tomorrow, and we slip into the worst-case scenario through repeated inaction. I think of these as Chronic Crises.

To my mind, the quintessential Pure Crisis is when there is a clear stable door and one can, after the event, see which side of it the horse is on and who acted or failed to act to shut it. Many of the Chronic Crises we slip into resemble a well-stocked stable losing horses day in, day out, while everybody thinks someone will get around to shutting the doors tomorrow, but they don't.

Is the world in a wildlife crisis? There are many ways of looking at this at a global level, but three of the most illuminating are extinction rates, species declines and ecological intactness of areas.

We're said to be living in the sixth extinction crisis that has hit Earth, or maybe the seventh depending on how you count them. But even that count of six or seven is a bit dodgy since all those extinction crises have occurred in the last 600 million years of Earth's 4.5 billion years of existence and are based on records of largeish organisms that had bodies that could turn into fossils. It's quite possible that there have been many, many more extinction events as big as the ones we know about, but that we simply do not know about the earlier ones.

Life has been kicking around on this planet for around 3.7 billion years. Our knowledge of it is pretty rudimentary for the first 3 billion years or so. Since our species has been in existence for a mere 300,000 years, and the invention of the microscope and telescope only happened about 450 years ago, I don't think we should beat ourselves up too much about being a bit hazy about the first few billions of years of life on Earth and not completely sure about what happened in 'recent times', say the last 550 million years since the start of the Cambrian Age.

The fossil record shows baseline rates of extinction at around one in a million species going extinct each year, but nowadays the rate is estimated to be maybe hundreds and quite possibly thousands of times higher than that. I'm slightly sceptical about that as there may be roughly 8 million species on Earth so we should expect about eight a year to disappear at the baseline rate but between 800 and 8,000 at the estimated higher rate of nowadays. So, can anyone tell me what the 50,000–500,000 species are that have been lost from the planet in my lifetime? The extinction rates and numbers of species on Earth are both pretty dodgy numbers in these calculations – only my age is known with quite stunning accuracy (to the hour, in fact). I don't doubt that extinction rates are much higher now than in the remote past, and we understand pretty well why that is. However, it is also the case that there are more species alive on Earth now than at any other time and that the current extinction crisis has hardly made a dent in numbers because it is only just starting. But give it a few hundred thousand years and it will show up on the graphs.

The Living Planet Index (LPI, produced by WWF) attempts to collate the population trends of species and blends the trends together to make an index of what is happening now. Because data are scarce, particularly for plants and invertebrates and the marine environment, the index is deliberately selective about using data from amphibians, reptiles, fish, birds and mammals and plots an index from 1970 to recent times (2016 is the cut-off for the LPI published in 2020). As more data are found, and more studies come into their own, then more data will populate the LPI and for a wider range of species, but already the LPI has fallen from a nominal starting point of 1 to a current average world value of 0.32 (in other words, a 68% fall in 46 years, the period since I took my A-Levels). On the face of it, we are told that terrestrial vertebrates have declined by two-thirds in abundance in far less than the average human lifetime. That takes into account the species that are increasing as well as those that are decreasing – these are the average declines.

The LPI can be looked at regionally too, and, as always, a geographic breakdown is very informative. That –68% globally represents the combination of five figures, all big declines, in these five parts of the world: Europe and Central Asia –24%; North America –33%; Middle East, South Asia and Australasia –45%; Africa –65%; South America –94%. Those are staggering figures, and whilst I don't believe that the average vertebrate population in South America has fallen by well over 90% in almost 50 years (do you?) I do believe that this gives a good indication of where the biggest wildlife losses are occurring.

The Biodiversity Intactness Index (BII, produced by the Natural History Museum) comes from a different approach to look at the same issues. Here the timescale is much longer, and although species trends are being built into the index, so are measures of habitat cover to represent the intactness of the regions compared with their natural ecosystems. The BII too is work in progress, finding new sources of data to populate the index, but its current embodiment plots a graph of biodiversity intactness for the last thousand years for the planet and for a variety of regions of the globe.

The BII for planet Earth was at around 98% intact in the year 900 and remained above 96% until around 1700 when it took a more rapid downward path, reaching 94% in 1800; then things got pro-gressively worse for the next two centuries, dumping us at a BII of just under 80% in 2000. This suggests that biological intactness has declined by 20% overall in a thousand years, with most of that loss occurring in the last 200 years. The last century of loss has been the worst ever, and the global index is still heading downwards. If this is a crisis, it is an accelerating Chronic Crisis.

When we look at a geographic breakdown of the BII then the striking thing, to me at least, is that areas such as North America and Western Europe which have experienced the largest drop are now slowly increasing their BII, whereas those parts of the world which have maintained their wildlife the best over the centuries, such as central Africa and South America, are now losing it at the fastest rate. A reasonable way of looking at this picture might be that in the rich north of Europe and much of North America we trashed our wildlife first and are now slowly putting some of it back again, while at the same time we are moving on to trash the most pristine parts of the remaining natural ecosystems on Earth.

In Western and Central Europe, the BII is very low, 64%, but that is a recovery from a low of 58% in the 1940s. We are told that the UK is one of the most nature-depleted countries on Earth and sometimes that it is the twelfth-lowest of all countries and territories for biological intactness. This is quite likely to be true, given that we cut down our forests hundreds and in some cases thousands of years ago, we don't have many high mountains that are somewhat immune from development, and we have concreted over about 8% of our land area. To climb far up the list, we'd have to plant much of the country with trees and maybe knock down a few built-up areas and let the forest come back. There are some practical difficulties in doing any of that, and its consequences for human quality of life on a small island with 68 million people would be extreme.

There is a worrying message from these figures, and that is that many of the countries which top the list of wealthy nations, with the highest per capita GDP, are also those with the highest loss of BII. If you lived in Brazil or the Democratic Republic of the Congo you might argue that the rich north has built its wealth on the back of ecological destruction at home. It's an uncomfortable truth that many of the countries which have experienced the biggest loss of wildlife intactness are rich and those that have the highest wildlife intactness are far further down the list of wealthy nations. What sort of message is that?

Is the world in a wildlife crisis? Yes, we are, and it is very much a Chronic Crisis rather than a Pure Crisis, but a crisis it is all the same. As a planet we are in a period of mass extinction, with continuing long-term and steep short-term decline in abundance of species.

The House Martin effect

Anyone over the age of, say, 40 years will have personal experience of changes in wildlife abundance during their lifetime. They will be, like mine, mixtures of gains and losses, and mixtures of things that seem explicable and inexplicable. For most of history we have had to extrapolate from our own observations and make intelligent (or wild) guesses at what is going on. Two great naturalists, Gilbert White and Charles Darwin, were keen observers of wildlife around them and corresponded prodigiously with others to try to assess whether or not what they had noticed was also happening elsewhere.

Nowadays we can search the internet for evidence, and we may belong to organisations which collect evidence of population changes and investigate their causes and publish the results. But we are still driven to a large extent by our experience. When I worked at the RSPB then often, as spring progressed, there would be conversations along the lines of 'Has anyone else noticed fewer House Martins this year?' – followed by lots of us basically saying 'yes' or 'no'. There was usually a range of views, and I assume that we were all perfect observers, so that range of views indicated a truly patchy pattern out in the real world.

I pick on the House Martin deliberately because it is a real example from the past. And it is a real example from my own life. In the two streets in Raunds in which I've lived over the past 35 years, House Martins have stopped nesting on the two houses where there used to be multiple nests, and they disappeared a few years back now. And on a farm I know well, also in Northamptonshire, the House Martins that used to delight and annoy (with their droppings) ceased to nest a few years back. But at the time of the water-cooler type conversations at the RSPB I used to say that the long-term survey evidence suggested a decline but nothing much to get too worried about yet.

Now, looking at those same survey data years later, it is estimated that in England we have lost almost three-quarters of our House Martins over the last 50 years, although the sample sizes of the early years are low and so that very long-term trend is a bit unreliable. But it does look as though those worried about their local declines were right, there was something going on. On the other hand, I assume that those people saying that nothing was happening where they lived were also right at that time – it was just that whatever the cause of the House Martin decline was, it had not reached them yet.

If we look carefully at the data from more recent years (from 1995 to 2017) we find that the overall UK decline is only (!) 18% in that period. The loss of almost one in five House Martins is quite a lot of House Martins from quite a lot of streets, but other species have declined more in that period (and some others have increased a lot more too). The picture becomes more interesting still if we look at the breakdown of that 18% across the four UK nations: England –34%, Wales –12%, Northern Ireland +98%, Scotland +120%. So, if an English woman, a Welsh man, a Northern Irish woman and a Scottish man walked into a bar and talked about House Martin declines, and they were all in their thirties and had good knowledge of what had happened in their own countries in the past twenty-something years, then they would all tell a different tale and they would all be right. The Englishman twice their age who joined the conversation might say 'But the decline you've seen in England since the mid-1990s is nothing compared to the decline I saw from the 1980s to the mid-1990s – that's when the damage was really done.' And he might add, 'But it is really good news that House Martins seem to have increased in Northern Ireland and Scotland. I didn't know that. I wonder why that is.'

There are plenty of consequences that flow from this little case study of how we know what we know about wildlife population trends. First, even if our memories are perfect (they aren't) and our local observations are correct (mine are but I can't vouch for yours), it is only when they are brought together to form a coherent whole that we can really have confidence that we know what is happening. Second, that requires a means of collating observations collected in a somewhat standardised format and analysing them as a whole – that's what monitoring schemes such as the BBS do and also, to a lesser but still to a large extent, what citizen-science projects such as the BGBW do. Third, there are always indications of decline somewhere or other in some species or other and some of these turn out to be early warning signals of future much more widespread declines – but many don't. Fourth, the population trend of a species doesn't really tell you much about its cause although the variations in trend might be very useful. So imagine you were sitting in England thinking that House Martins had declined by three-quarters in your lifetime. You might run through some potential causes: loss of aerial insects (maybe caused by agricultural insecticides or general air pollution); loss of nest sites caused by people poking out House Martin nests with broom handles; disease; predation from the burgeoning Hobby population (or maybe from Grey Squirrels); something affecting them on their wintering grounds in Africa or during migration, or in some way from climate change. But the differing trends in England, Wales, Northern Ireland and Scotland might make you lean more strongly towards one or other of those as the thing it is most worth investigating further.

We are almost always dealing with chronic long-term declines; loss of wildlife is an ongoing, gradual but it appears inexorable thing. If the House Martin has declined by 72% in England since 1966 then by my reckoning only 23 of the 51 successive years (the published graphs I am looking at take us to 2017) have been years of decline, and there have actually been more years of slight increase than of decline. But the declines have been, on average, bigger (quite a lot bigger). Taking the English data at face value, there were two periods over which the decline mainly took place: the first in the decade of the 1980s and the second from approximately 2005 onwards, with 1990–2005 being a period of stability.

If only House Martins had got their act together and decided to decline by 72% in a single year, that would have made a massive impact. But they did it all wrong if they wanted anyone to take any notice of it. And those long-drawn-out stuttering declines, sometimes with false signs of recovery, are typical of the losses of wildlife we have experienced in the UK.

Wildlife loss is a story of chronic decline in the UK these days. That's actually what most changes are like, whether they be social or environmental. Both good things and bad things happen slowly and gradually. But many small changes add up to big changes – and the lessons from that are, I suggest, that we can expect gradual progress in the things that we get right but we should not become inured to gradual decline in the things that we get wrong, because those gradual small declines add up to massive losses over time.

The horsemen of the ecological apocalypse

There are a lot of species out there, and even as someone keen on wildlife, I would be hard pressed to name many of the 70,000+ species that inhabit the UK. I could do the birds, mammals, reptiles and amphibians pretty well, and get through quite a few fish, but that wouldn't take me to my first thousand species. I could do a fair few trees and shrubs and lots of plants but not a high proportion of them, and then I'd do well on butterflies but generally poorly on all invertebrates and appallingly badly on most other groups, particularly in the marine environment. For someone who cares about wildlife, I can't name many of the players – but then I care about people too and I couldn't name anything like a million of the UK's 68 million people. It doesn't mean I don't care.

One can see why a politician or journalist dropped into this area through their job, rather than through their inclination, might blanch at the number of species. If you don't know your Snake's Head Fritillary from your High Brown Fritillary, or your Adder from your Adder's-tongue Fern then the natural world must seem confusing.

But for those of us who are mesmerised by the beauty of the natural world, its complexity is part of the attraction. To balk at the number of species is a bit like the interchange between Emperor Joseph II and Mozart over his music in the film *Amadeus*, where Mozart was told that his music was very good but there are just 'too many notes' and that he should cut a few and it would be fine. You can't criticise Mozart for 'too many notes' – it's the notes, and their abundance and combinations, that make Mozart what he was and is. And that's how we should feel about wildlife too.

Luckily, though, when it comes to understanding the drivers of decline, the picture is far simpler. Species don't go extinct, or decline in numbers, on their own very often. The cause of wildlife loss is almost always some challenge to the system, some assault mounted by our species, sometimes deliberately but usually carelessly and thoughtlessly. And those assaults, whether global or national, are of four main classes, which we can think of as the four horsemen of the ecological apocalypse. Instead of the biblical War, Famine, Pestilence and Death, our horsemen are Invasives, Overharvesting, Pollution and Habitat Loss.

Invasive non-native species are a common cause of global extinction, and a common cause of UK wildlife loss. The non-native species could be a predator which munches its way through a native species (such as American Mink and Water Voles) or it may be a disease spreader (such as Grey Squirrel spreading diseases to Red Squirrels), or its impacts may affect whole habitats (such as Muntjac Deer browsing the shrub layer in woods to the detriment of ground- and shrub-nesting birds, perhaps including Nightingales). And often, the impacts are multiple and sometimes not that well known. We mustn't ignore non-native plants either, for they cause some of the biggest impacts.

The number of UK native species driven to national extinction by non-native species is small, but the number much reduced in numbers is large.

One of the more ridiculous introductions of a problem species is that of the American Signal Crayfish into many European countries, often so that it can be hunted for food. The UK has only one native crayfish species, the White-clawed Crayfish, whereas there are several crayfish species in Europe. The Signal Crayfish carries a disease, a mould, to which it is largely immune, except when stressed, but all European crayfish are highly susceptible to the crayfish plague. In addition, the larger American species will eat our smaller native species. Since the Signal Crayfish were released here, not much more than 30 years ago, the native crayfish have disappeared from many previously occupied rivers, to be replaced by this American species. The same is happening in other European countries, and population declines amount to over 50% in a 10-year period.

Overexploitation can be a potent cause of wildlife loss. In the UK, overfishing of the seas is probably the easiest bunch of cases to point to, and it almost beggars belief that even though it is in the fishermen's own economic interest, and in the interests of the communities in which they are embedded, to take account of the sustainability of harvesting, everything still goes to pot. A few terrestrial species have been extirpated in the UK by overhunting, either because they were big and fierce and posed a threat, real or perceived, to people's lives or livelihoods (e.g. Brown Bear and other large carnivores) or because they were the subject of collecting for trophies or specimens (e.g. various bird species). But, generally speaking, overexploitation is a relatively small part of the problem in the UK. The decline of the Grey Partridge on British farmland is not because it is a gamebird – and indeed the loss would probably have been even greater had it not been a gamebird that some wanted to survive in numbers – but because its habitat has been massively damaged. I can't think of a British plant species that has been picked to national extinction, and if there is one, I'd be willing to bet that it got to that parlous state through some other factor.

The third horseman is pollution. Water pollution cleared out much of the wildlife from many of our rivers in the past, and this is an area where we have, as a country, moved slowly but quite effectively. Far fewer waterways, even in industrial areas, are now grossly polluted. A polluted waterway, particularly if it stinks, is obvious to the senses, but agricultural chemicals used in the countryside are largely invisible. The Sparrowhawks that were very rare sights as I grew up in Bristol in the 1960s are now much commoner because we have removed the chemical pollutants that affected their survival, their nesting success and their food supply – another success story. But we now worry about the impacts of agricultural products on insect populations. Nitrogen pollution, falling from the skies as a weak fertiliser, is another example of a potent pollutant which we tend not to notice in our daily lives but which has far-reaching impacts. And it's sometimes difficult fully to appreciate that an unseasonably warm and sunny winter day may well be one of a large number of events amounting to systemic climate change, affecting temperature, rainfall and storminess that affect not just our heating bills and flood risk but the ability of species to function as they have done, and of food webs to remain stitched together. Climate change is real, affects our wildlife now, and will come to dominate wildlife conservation more and more as time passes. Its impacts are already clear, and becoming clearer year on year, and they may come to dominate the wildlife reviews of our grandchildren's generation.

Looking back at what has happened, though, the mightiest, the most destructive of the four horsemen of the ecological apocalypse has been habitat loss and deterioration. The losses of ancient woodlands, meadows, chalk grasslands and lowland heaths have greatly reduced the abundance of the species that are somewhat specialised to live in those habitats, but even the fragments that remain won't be rich in wildlife unless they are in good nick. Habitat quality is the key to wildlife richness in the UK. We must protect the best of what is left, improve the quality of the degraded, and look to recreate more and better examples of what has been lost. And we mustn't get too hung up on what is a habitat or which habitat this is, or indeed whether it is a natural, semi-natural or artificial habitat – wildlife exists everywhere, and everywhere can contribute, though not necessarily equally, to a wildlife recovery. Most of the UK land area is farmed – and farmland is not a natural habitat, although we could stretch a point for some types of farmland and call them semi-natural habitats. But since farmland covers some 70% of our land area, and is the scene of some of the most dramatic wildlife declines, both over the long term and over the much shorter term of a human lifetime, farmland wildlife should be a big focus for wildlife recovery.

Four apocalyptic horsemen ride through the wildlife of the world and our seas, our forests and our farmland, and they must be defeated if we are to see wildlife thrive. Vanquishing them is the essence of wildlife conservation. How can we unhorse the horsemen?

Reflection 2

No-one living today has lived a life during which wildlife in the UK has increased in abundance. No-one. And nor did our parents or grandparents. When Albert Knighton lived in my home in 1899, and set off through the Victorian front door that is now mine, to work in the shoe works as a clicker, he was living through an earlier stage of the same chronic decline in wildlife that we are still experiencing, and so did his parents and grandparents. If you are a wildlife enthusiast it's a depressing background, which, on the face of it, is not encouraging for the future. But we might as well face facts.

My own baseline expectations of wildlife around me were set over 50 years ago in an area on the south side of Bristol where I grew up. The North Somerset countryside and my journeys into Bristol for school set my standards of normality. Water Voles were commonplace on the River Chew in my youth, but their numbers are now much reduced and when I revisit I feel their absence. The Spotted Flycatchers that I first saw on a May morning in the early 1970s have declined hugely in numbers but the Red Kites that I never saw in those skies now occur in places from which they have been absent for over a century. The adult that I am now had my expectations of wildlife abundance set by my childhood experiences, just as my expectations of fairness, decency, food and material comfort were set then. Now that I am an ageing campaigner, my aims are not to recapture the good old days, because they weren't uniformly good, but to try and create a better future, which might well involve recreating some of the best of the past, but also is founded on a belief that the future can be better than the past was, and than the present is, if only we make it happen.

A wildlife-savvy child growing up in Raunds today will be having their own baselines set for wildlife and for human society. Their streets are more ethnically diverse than they would have been for their grandparents, their bus services will be poorer, the food they eat will be more varied, and their skies will contain Red Kites every day, but Spotted Flycatchers will be rare sightings indeed. Hedgehogs seem almost mythical, in that parents or grandparents talk about them as regular garden visitors but many children will never have seen one alive.

The concept of shifting baselines is relevant here. In a world where a particular aspect of wildlife may well have declined we often consider that the earliest situation we can remember was 'normal' – but we all too often forget that what was normal then was regarded as a failure at the time. The shifting baseline concept is more about our ambitions for environmental improvement than merely a statement that things used to be better. How much more wildlife do we want? And because wildlife has declined, that is akin to asking how far back in the past would we like to go? .

Would I want to go back and live in Albert Knighton's time? Hell no! Because that would mean giving up many material comforts that I have, and my family and neighbours have, to go back to a time of long working hours, little job security, greater racism, no National Health Service, no unemployment pay and no state pension. I wouldn't swap those things for a few more Turtle Doves purring in the neighbour-hood. No way! I have tried to spend my life providing wildlife with more chances and more successes, but not at the expense of my fellow human beings. Yes, there are trade-offs to be made – but I don't want to go back to the good old days with lots of wildlife, I want to go forward to the good new days of lots of wildlife, while at the same time protecting many, not all, of the material comforts that we have now. I doubt whether Albert Knighton would hesitate to live in my world, but I would certainly turn down the option to live in his.

It's difficult to escape the thought that my prosperity, the house I live in and the society in which it sits, have been built on environmental destruction just as they were built on exploitation of the UK population and people across the globe. I wasn't responsible for that ecological destruction, just as I wasn't responsible for slavery, but I benefit from their legacies.

Although I am unhappy about the state of wildlife this can't be a very widespread feeling or else, surely, we would have done a better job in fixing it sooner and better. How do we do better in future? That must be the job of wildlife conservationists.

CHAPTER 3

What is wildlife conservation?

I f there's wildlife all around us (as there is), and it's in chronic
decline (as it is), and those reductions are driven by human actions
(as they are), then we have a choice about whether to accelerate,
halt or reverse those losses. It's up to us, as we are the main drivers
of change on this planet and we have it in our power to change the
world. Personally, I'd like to live in a country with more wildlife on
my doorstep and in the distant parts of the UK, and I've spent most
of my adult life as a wildlife conservationist engaged in that pursuit.

Here I'll set out what I mean by wildlife conservation, although
it's possible that others might disagree. What is wildlife conservation, and importantly, what isn't included in this enterprise? How is it
done, and who does it?

This thing called wildlife conservation

Not long ago, I came across someone quoting me as saying
something along the lines of 'The goal of wildlife conservation is
to stop rare species from going extinct, to stop uncommon species
from becoming rare, and to stop common species from becoming
uncommon.' That quote takes me back about 25 years, but despite
the distance I think it is accurate (and I'm touched that anyone
remembers anything I say). As far as it goes it isn't a bad way of
capturing some of the essence of wildlife conservation, and it's worth
noting that it focuses on outcomes rather than actions. As we have
seen, however, we wildlife conservationists haven't had careers of
extravagant success – yet.

Looking back on those words, there is something both accurate and sad about them, in that they seem to refer to a world where things are going badly wrong and our aim is to slow down the rate at which wrongness accumulates. As we have seen, it's hard to argue with that as an accurate reflection of where we are – wildlife conservation is a profession whose successes are often measured by the extent to which it slows down the rate at which harm accumulates. Need it be this way?

Even all those years ago, I don't think I would have signed up to that as a complete definition of what wildlife conservation's objectives should be. If we see wildlife conservation as a brake on activities that lead to species decline then we have been on that downward slope for many years, and so it is perfectly reasonable – indeed, I'd say essential – to try to move back uphill in some areas as well as digging our heels in to try to stop the overall decline. Wildlife conservation is a restorative as well as a protectionist agenda, despite there being an awful lot of effort put into stopping the rot.

Conservation is a tricky word, meaning, as it does usually, keeping things much the same. Conserves are a culinary way of preserving fruits with sugar, grass conservation for a farmer means keeping quantities of forage from the growth season so that it can be fed to livestock in the winter months, landscape conservation is a matter of protecting the current view from harm, architectural conservation is quite specifically about prolonging the best of the past built environment through intervention, and the Conservative Party takes pride in heritage and tradition and perhaps too often looks to the past rather than to a brighter future. Wildlife conservation sounds like a pickling process in which some of the best of our wildlife is artificially kept as it is, and that is, to many of us, lacking the right degree of ambition.

Those connotations of the word conservation are unfortunate, but they shouldn't blind us to the fact that in an era of net loss of wildlife then protecting what remains is a crucial part of the job in hand. Preventing the loss of wildlife-rich sites, retaining areas of wildlife-rich habitats and stopping further population declines are fundamental aspects of the wildlife conservation programme.

However, we shouldn't, and we don't, stop there. Species reintroductions, habitat restoration, nature regeneration and ecosystem rewilding are all part of the mix too – they always have been but are an even more prominent part of wildlife conservation these days.

George Monbiot's book *Feral* was published in 2013 and had an amazing positive impact on public perceptions of what was possible. It raised the subject of rewilding, a word that means different things to different people but is essentially the restoration of more natural ecosystems. That might include such things as better management of existing habitat that has fallen into disrepair, the restoration of a once-present habitat that was lost to another land use, the reintroduction of species that used to occur on a site within living memory, or the restoration of a long-lost species such as Beaver or Lynx.

George's book was just a bit unfair to the work of wildlife conservation organisations that had been doing restorative wildlife conservation almost forever. However, the agenda has significantly shifted since 2013, with rewilding becoming much more talked about, much more mainstream and a much trendier part of the wildlife conservation mix.

Wildlife conservation is about stopping things getting worse, but it is also about making things better – and it always has been.

Does it matter?

Does the loss of wildlife from our lives really matter? Surely there's plenty of it left! And in any case, who needs wildlife anyway?

Such a nihilistic view of life can be asked of any human activity, from waging war to holding a door open for someone. The question really is, what sort of world do we want to live in? I am passionately keen to live in a world with more wildlife, and it doesn't really need justifying any more than that. After all, society accepts that listening to opera, and even to jazz, and watching people play football, and cricket, and seeing the art of Monet, and of Tracey Emin, are parts of our collective culture. They are legitimate things to want. I think that's the perfect answer to the question – it matters to me and many others.

There is a different way, a more self-centred way, of answering the same question, and that is in terms of us needing to stop destroying wildlife because otherwise we ourselves will suffer. There is something in this view of life, and it's sometimes told in the form of the 'rivets in the plane' story. Yes, we can afford to lose some wildlife, some whole species, some large areas of natural habitat, but it's rather like a plane in which you are a passenger – it can lose some rivets from the wings with nothing bad happening, but eventually the loss of too many rivets will lead to disaster. It's an appealing story, and I always think of it on the very rare occasions I am taking a flight, but the trouble is that there are probably an awful lot of rivets still to pop before we crash and burn, so it doesn't move things forward very much.

When I think rivets, I think Passenger Pigeons, which lived in the forests of eastern North America until about 1900. This was the most numerous bird in the world 50 years before it went extinct. They numbered in their billions, so you'd think that the Passenger Pigeon would be a sizeable rivet, but its absence has made very little difference to the ecology of North America and there is no phrase in common usage in Biden's America along the lines of 'If only we had Passenger Pigeons, life would be so much better.'

That's the trouble – wildlife loss is a Chronic Crisis. We're not all going to suffer dreadfully tomorrow if we don't stop the continued loss of wildlife today. We've been suffering a loss of wildlife in the UK for very much longer than a century and yet we are all more comfortable than ever in our daily lives. Many will take quite a lot of persuasion that there is an urgent need to be much better at protecting UK wildlife because we will face dire consequences if we push wildlife numbers further down.

And they are right to be sceptical. If I were to lament the loss of the Turtle Dove from the Northamptonshire countryside it would be perfectly fair for someone to say that they have never heard of a Turtle Dove (except, are they the same ones in the song 'The 12 days of Christmas'?) and have certainly never knowingly heard an actual Turtle Dove, and that they have made it through their whole lives so far without feeling any loss. If they were really clued up they could deploy the Passenger Pigeon case and say that if the USA can cope perfectly well without Passenger Pigeons, and it seems they do, then we can probably cope perfectly well without Turtle Doves. Indeed, if I then told them that Turtle Doves have declined because of our intensive agricultural systems they might respond 'What? That agriculture that produces cheap food. Sounds to me as though there might be a cost to me if we adjusted farming to allow the Turtle Doves back at the expense of production' – and in general terms they might well be right.

And yet, the Red Kite has been returned to our skies and we feel better for it. Most people did not feel awful when it was absent, and nobody died because it was absent, but many feel better now that it is back. They will feel better if their nearest Pasqueflowers are more accessible, if the Nightingale sings again in more woods and if Hedgehogs snuffle around in more gardens.

Wildlife is fascinating and beautiful – and it's real. There is nowhere else in the universe that has Herb-Robert, save my street and other places on Planet Earth. Noticing wildlife is part of showing that you are an inhabitant of this planet with the ability to look up from the world of human soap operas, the state of stock markets, and the football results to notice that there is a world wrapped around us that would exist without us, and is doing its best to withstand the insults and challenges that we throw its way. There is nothing more special about Earth, our home, than the Blue Whale, the Herb-Robert, the Black Garden Ant and herds of Wildebeest.

Of course, there is something in the rivets in the plane idea, but the analogy applies more to protecting things at the scale of the Amazon rainforest than the Herb-Robert flower that gives me so much pleasure or the number of House Sparrows on my feeders. After all, we claim to live in one of the most nature-depleted countries in the world, and at a time when that depletion has been very high, and yet when we look around the streets of our towns, the people living in them, on average, have a far better standard of living than their great grandparents. We live longer and generally have an easier life. The rivets have been popping like mad and yet it is difficult to argue that our comfort is about to take a very sharp turn downwards because of the loss of wildlife. It might be easier to argue the case that our comfort is built on wildlife losses rather than imperilled by them. Wildlife conservationists have to be careful not to exaggerate the impacts on us all of the loss of wildlife. Wanting more wildlife is a perfectly reasonable desire for us to have individually, and one which society as a whole might well share, but we'd be foolish to assume that everyone desperately needs more wildlife.

Sustainable development

Progress is generally marked by what we leave behind as much as by what we acquire. My life has seen huge progress on many issues: workers' rights, more equal rights for women and ethnic minorities, improvements in pensions, health care, holidays, education and social mobility. There is more to be done, and we must do it, but it's worth celebrating the journey so far.

In stark contrast, wildlife has taken a complete clobbering over the same period. Why is it that we can point to economic and social improvement but not so easily to environmental and wildlife success stories? We've done better for our fellow man (and woman) than for our fellow species.

There was a time when we all talked a lot about sustainable development as meeting the needs of the present without compromising the ability of future generations to meet their needs. Sustainable development was like a three-legged stool where the economy, societal fairness and the environment were the legs. Unless the three legs were of acceptably similar length we would be in an uncomfortable position. You don't hear so much about sustainable development now, but it remains a handy way of looking at the world, and at wildlife conservation's place in the bigger picture of human endeavour.

One reason we hear less about sustainable development is that the Westminster government, elected in May 2010, announced in July 2010 the abolition of the Sustainable Development Commission, a UK body set up in 2000 to advise governments across the UK. The SDC was one of the few places where the tensions, sometimes stark conflicts, between economic progress and social and environmental progress would ever be discussed. The profit motive is so strong that social equity and environmental sustainability both need privileged seats at the table, otherwise they won't get a look in when decisions are made. Perhaps the last thing that the new government wanted was a body pointing out inconvenient truths about the impact of poorly restrained economic growth on the environment.

In Wales, the Welsh government has tried to put sustainable development at the heart of its decision making in a way that decisions in England and the work of the UK government do not. Wales has a Future Generations Commissioner to take the long view and build in the needs and wellbeing of those who do not have a say because they are yet to be born. Sustainable development has been made the single organising principle for all public bodies in Wales, and this approach may be making a difference.

A major (£1.5 billion) transport infrastructure project, known as the M4 relief road, or the Newport bypass, or concreting over the Gwent Levels, was rejected by the Welsh government in June 2019, after long deliberation. The reasons given were a combination of cost and environmental impact. The First Minister of Wales, Mark Drakeford, said 'I attach very significant weight to the fact that the project would have a substantial adverse impact on the Gwent Levels Sites of Special Scientific Interest (SSSIs) and their reen [drainage ditch] network and wildlife, and on other species, and a permanent adverse impact on the historic landscape of the Gwent Levels.' Now you might consider that the environment was being used as an excuse here, but that wasn't necessarily so. I will give the Welsh government the benefit of any doubt that exists and say that they really did weigh up all three legs of sustainable development and came to a rounded decision. We need more such decisions. No such detailed assessment has been made for the High Speed 2 (HS2) rail project in England, which has a 50-times higher financial cost than the Welsh road scheme, and arguably a much higher social and environmental cost too.

Thinking in the sustainable development way, of impacts on the economy, society and the environment, won't always generate win-win-win outcomes, but it encourages us to look for them. Also, it tends to alert decision makers that many projects which they thought would create large economic gains at little environmental cost have turned out to be less good for the economy than was promised and more environmentally damaging than was feared. Those are useful lessons to learn.

Wildlife conservation and climate change action

People often now talk about the climate and wildlife crises. That seems to me to be a sign of real progress and a sign of hope – talking isn't the same as doing, but it is a useful start. Sometimes I hear talk of the wildlife and climate change crisis – crisis singular, as though these two things are either the same thing or are so joined at the hip that they cannot but be seen as one entity. This seems to me to be both wrong and unhelpful. For sure, the loss of wildlife and the increase in greenhouse gases are caused by a similar suite of human actions which could be summed up as material progress, but they aren't exactly the same and solving one certainly won't solve the other unless it is by a fluke. The reintroduction of the Red Kite is a small but significant wildlife success story, but unless I've missed something, I don't think it will have cut down greenhouse gas emissions.

A new phrase, or piece of jargon, is 'nature-based solutions' to climate change. These are measures that involve the natural world which will either reduce the amount of climate change which happens or reduce the impacts of those changes on us. Increasing forest cover is certainly one of the former cases and is often one of the latter as well. Whereas building lots of windfarms isn't a nature-based solution to climate change, 'building' lots of forests is, and whereas 'planting' lots of wind turbines sounds silly, planting lots of trees doesn't sound daft. The words we use help us here. And just occasionally, a landowner might well be choosing between those two options as a land use on a piece of ground – choosing between planting wind turbines and building forests. Which is the right answer will depend on what you want to achieve because for any particular site the reduction of greenhouse gas emissions from either option will differ, and the impacts on the amount of wildlife on the site will also differ. Depending on their feelings about wildlife losses and greenhouse gas increases, different people might choose different options. And the costs and benefits of using the land to build a windfarm and a solar farm will be different too, not only in financial terms but also in terms of wildlife profits and losses and greenhouse gas profits and losses, just as the options of planting fast-growing conifers, slow-growing broadleaf trees or allowing a site to scrub up and eventually turn into a woodland will vary too.

It goes without saying that the wider environmental issues and the wildlife issues aren't necessarily at the very top of the landowner's mind when she (or he) comes to decide what to do with the land. They may well be interested in the long-term financial outlook from trees versus wind turbines, conifers versus broadleaves, and natural regeneration versus planting, and the short-term financial outlook will play a big part too. Actually, the landowner's mind might be full of thoughts of getting out of, say, unprofitable sheep farming rather than anything to do with the environment and/or wildlife. And the views of the local people will be driven by thoughts of the impacts of climate change and wildlife, but also access to the land, the landscape, how many jobs will be created and will they go to locals or will a workforce arrive, the amount of traffic on the roads in construction or planting times and whether the forestry or windfarm company involved will sponsor the local football team. In practice, and perfectly reasonably, different people will approach any issue with a bundle of views, and those views will be shaped by what they think is important. And importance will be judged across many different values and currencies. The point is, we won't all be starting from the same place. Wildlife conservation and environmental progress are close neighbours but aren't exactly in the same place.

I've never heard anyone calling a nuclear power station a nature-based solution, and the environmental movement has been pretty uniformly against nuclear power for ever. I'm not entirely sure why. It's partly a matter of personal development: some of the environmentalists of the 1970s and 1980s were very much from a CND, Ban the Bomb, background and their values had been shaped in the post-Hiroshima age. Nuclear power was seen as dangerous on many different levels – unsafe to operate and unsafe to have that technology around to fall into the wrong hands. Those arguments have weight but they aren't environmental arguments. At the RSPB, as we developed our thinking on climate change and its impacts on wildlife we also had to develop our thinking on climate mitigation – what measures society would need to put in place to limit emissions. And nuclear power was clearly one of the options. When we, staff, took a policy paper on climate mitigation to RSPB Council we spelled out the ramifications of climate change for the natural world and discussed the options for reducing those impacts and recommended that nuclear power was one of them. So we ended up in a pro-nuclear position. This was one of the few decided RSPB policies that we didn't aim to use very much. And we didn't aim to use it because we didn't want to get dragged into a lot of difficult debates which were on the edge of our expertise, and the edge of our interests, and which we probably couldn't influence very much anyway. And we didn't want to fall out with our fellow NGO friends over these matters either.

Our position was partly based on the thought that if we are to reduce greenhouse gas emissions then we will need a wide variety of potential solutions to deploy at different times and in different places, and that the footprint of nuclear power stations is very small compared with generating all that power from wind turbines, solar farms and other methods, and much better than using coal or gas. So our position was that nuclear was part of the mix and we said that, although not stridently and not in ways that were designed to stir things up, but we certainly said it.

Greenpeace came from a different history and a different position and would never – that is, never at that time – have agreed to nuclear power being a valuable part of the mitigation toolkit. But as the importance of climate change became more and more mainstream and recognised then it became more and more difficult to rule out nuclear power as an option. If you like, and if you agree that it is an option, it was easier for RSPB to waltz in to this issue late in the day and see that, than it was for organisations who had been embroiled in similar issues for a very long time to change their position. But it does show, I think, that environment and wildlife aren't the same thing at all. To me, the difference is self-evident, and although they often go hand in hand, that isn't always the case.

Although a nuclear power station is not a nature-based solution to climate change it might be a nature-friendly part of the solution. If you believe that climate change is real and caused by human actions (as I do), and if you believe that we can and should do something about it for many reasons (as I do), and if you believe that power generation must play a large part in that (as I do), and if you believe that nuclear power is a potent and viable option as part of the mix (as I do), then building nuclear power stations is something that you might support – and I do. You may depart from that line of reasoning at any of the 'ifs' along the way, but I don't. And so the question is where to stick the nuclear power stations. Location is always important in such decisions, and I see that the RSPB has now got into an argument about a third nuclear power station, Sizewell C, near its flagship wildlife reserve of Minsmere on the Suffolk coast, thereby opening itself up to a charge of Nimbyism (not in my backyard-ism). I have deliberately not looked hard at the arguments flying around on this subject but I guess they will involve some difficult balancing of local and national interests, and some of them will probably be about how it's done as well as whether it's done, and what the mitigation package should be.

There are many people who would call themselves both environmentalists and wildlife conservationists – I am one of them. But I know of some who are wildlife conservationists but climate change sceptics, and a few whose main focus is climate change who frankly don't give a stuff about wildlife. The two perspectives are not the same.

Wildlife conservation and animal welfare

How does the welfare of individual animals fit into this picture? And, even, should it? Well, the answer to a complex question such as this is often yes and no, and so, I believe, it is here. Animal welfare issues are not conservation issues, but they are important issues. Let's take the Badger cull as an example. Badgers are being culled by government order across large parts of England, and the area is widening all the time, in order, in theory, to limit or even indeed eradicate bovine tuberculosis in dairy cattle. Leaving aside whether or not this will work as a disease control mechanism, this programme of culling is leading to great reductions in the Badger population (an otherwise protected species) and that is a conservation issue. At the same time, a large part of the cull is being done by shooting in the dark at free-living Badgers rather than trapping them and then killing them quickly, and that is an animal welfare issue. Now I suspect that most of us animal lovers, as I'd be happy to be called, care about both aspects of the cull – but they are separate. I might oppose the cull even if the Badger population were relatively unaffected, if individual animals were to suffer horrendous pain and suffering; and it would be possible to support the cull even if the Badger population were to fall massively, if one believed that no unacceptable pain and suffering were being imposed on Badgers. Indeed, I guess the latter is the Westminster government position on this matter – they want fewer Badgers, and they would argue that the instances of suffering for Badgers are few and an acceptable price to pay.

The Badger cull is an economic issue in which the welfare and conservation concerns are both involved on the same side, but there are also issues where it is the very conservationists and animal welfarists (for want of a better word) who are in opposition. If it were possible to eliminate Grey Squirrels in a locality, and if that led to an increase in Red Squirrel numbers, would that be acceptable? To some, I think many, conservationists it would be. There would be discussions about the humanity of the killing method, the effectiveness of the killing as a means of reducing Grey Squirrel numbers and as a measure to benefit Red Squirrels, and a host of other issues but it is easy to imagine that many conservationists, once those concerns were allayed, might well sign up to such a plan. But there would be others, more concerned about the animal welfare issues, who would not, and who would hold the view that even a non-native Grey Squirrel deserves our consideration and that such a creature has a right to life that is not trumped by any benefits to Red Squirrels that might accrue.

That's a somewhat, although not wholly, hypothetical example – but an absolutely real one is whether to cull (i.e. kill) some reasonably numerous predators (e.g. Red Foxes) in order to protect other creatures that are somewhat endangered (e.g. breeding waders such as Lapwing and Curlew). Many see fox killing for wildlife conservation as being as bad as fox killing for fun, and also seem to see the limited and targeted (and sometimes experimental) fox killing of conservation organisations as being on a par with the widespread and routine fox killing which occurs on much land that is managed by gamekeepers. Now I do not seek to tell you what you should think about this, but I'm happy to expose my own views to scrutiny.

Do I want Red Foxes to be killed? No, I'd rather they weren't if possible. Would I put up with some Red Foxes being killed? Yes, I would, if that were done humanely and if it had a real conservation outcome, and if as few Red Foxes were killed as possible. And I would prefer it if Red Fox impacts on species of conservation concern could be reduced by non-lethal means such as fencing rather than shooting. I guess my position is a middle way between blanket fox control (i.e. killing) as practised on shooting estates (and not primarily for a wildlife conservation benefit but to protect game which is there, on that land, to be shot at by people) and an absolute belief that no wild animal should ever be killed by humans. It certainly feels like a middle ground, because if one adopts it one is attacked from both sides. The shooting community will say that you are hypocritical because you aren't doing your best for wildlife conservation and the welfarists will say that you are hypocritical because you profess to care about wildlife and yet you are killing some of it. Clearly the welfarists and the shooters don't agree at all with each other. On a scale from 0 to 100, most pro-shooting organisations seem to be nestled in the 90–100 range whereas most land-owning conservation organisations (the National Trust, the large number of individual Wildlife Trusts and the RSPB primarily) are down the other end of the spectrum in the 0–20 range as best I can tell. Killing wildlife for wildlife conservation gain really is a last resort for most conservation organisations, and I think that is how it should be.

Let's just imagine, as a hypothetical example – another thought experiment – that the RSPCA very occasionally uses rat poison in some of its animal sanctuaries if it has a problem that cannot be solved in any other way. I could sympathise, although the use of poisons is something at which I would certainly balk and put a lot of effort into avoiding – but I guess that the RSPCA would too. The occasional use of poison might push the RSPCA up as far as point 1 or 2 on the scale of 0–100 referred to above. How would we feel if an organisation that puts its whole being into maintaining the welfare of animals imposed suffering, on, say, 100 Brown Rats a year? I'm not attempting to tell you what you should think, but my point is that if you were to lose some respect for the RSPCA under those hypothetical circumstances then I think that would be understandable, but if you lost all respect for the organisation then I think you'd be throwing babies out with bathwater.

This is a complex and fraught area, but I think is clear that although wildlife conservation and animal welfare can both be shoved under the same heading of 'being nice to wildlife' they have different value systems. They usually rub along very well, but there are some rather limited circumstances where they clash.

What conservationists do

There are many ways of doing wildlife conservation, and each has its strong proponents.

It may be helpful to consider the 70,000 species in the UK as being akin to the population of a town the size of Bognor Regis, Stafford or Carlisle with 70,000 inhabitants. If the conservation agencies and charities intend to keep all these species in a fit and healthy state then it would be helpful if each of them reported for a regular check-up so that its needs could be assessed. We've seen that that is far from the case. However, enough cases present themselves to conservation bodies as worthy causes deserving assistance for the demand for conservation care to outstrip supply, and some sort of formal or ad hoc triage system has to be in place. It's just not possible to get involved with every declining or threatened species, and so some judgement has to be made, usually on the basis of cost and likelihood of success.

For conservation interventions to work we have to have a reasonably good idea of the cause of decline, so that we can implement the right solutions. There is no point reintroducing a species to a habitat if it is the poor quality of the habitat that is causing the decline in the first place – it's not going to lead to recovery any more than prescribing aspirin will mend a broken leg. Sometimes there is lack of clarity over the cause of a decline, and that may be because it is difficult to study or perhaps because there really are multiple causes for the plight of the species. One example is the Turtle Dove that I now rarely see at my local patch of Stanwick Lakes. Has it disappeared from there and many other sites because Turtle Doves are shot in unsustainable numbers, because of climate change affecting them in some way, because of a disease, or because farmland now doesn't meet the needs of this bird as it once did? In other words, can we identify the responsible horseman of the ecological apocalypse? In this case I'm not sure, but my money would be on disease or habitat deterioration – so that's where I'd put my efforts. Many would say that addressing any threat to the species would help in conserving a threatened species, and they would be right, but it's clearly better to solve the big problems than the small problems if we can. So diagnosis is important.

However, we don't save species one at a time, at least not always. In many situations, a conservation action addresses the needs of many species at once. Two broad categories of these might be land ownership and management, and shaping public policy to create a better world for wildlife. Protecting particular places by land ownership by a sympathetic individual or organisation can protect and help all the species in that area if done well. Likewise, the introduction of policies that raise the overall standards of agricultural management across the whole of the farmed landscape have great power to make a widespread difference to large areas and many species at a stroke of a policy-maker's pen.

Wildlife conservation is a business redolent with choices. Should we help Species A or Species B, and if Species A then should we implement Measure A or Measure B? Should we protect Site A or Site B, or should we be targeting Policy A or Policy B for biggest impact? These choices resemble many investment decisions and are influenced by the same thoughts. How much will this approach cost, what might we gain, and just how likely are we to realise that gain? Government and its agencies can involve themselves across all approaches, and it would be nice to think that there is some sort of overall strategy guiding investment in each. Environmental organisations tend to be constrained by their size to adopt particular approaches, but the bigger beasts in the wildlife conservation and environment sector appear to have made choices over which approach best suits them. For instance, Friends of the Earth operates at a policy level and not as a landowner, whereas the National Trust adopts almost a mirror-image approach. Some large NGOs, particularly RSPB, the Wildlife Trusts and the Woodland Trust, have long played a mixed game, or have, in investment terms, a balanced portfolio of actions across the risk and reward spectrum.

Who are the wildlife conservationists?

Are we all conservationists now? There is a much-talked-of biodiversity crisis and we all watch wildlife programmes on the television and feel a pang of discomfort in the bit near the end where we learn that things aren't going well for wildlife out there in the real world. However, it is difficult for most of us to buy and manage a wildlife reserve, or draft and then advocate for a new law to protect wildlife – so although we are wildlife supporters, we aren't really wildlife conservationists. We all rely on the experts, the professionals, to do most of our wildlife conservation for us. We pay our taxes and hope that governments, and their agencies, will do wildlife conservation well using some of our money. And we choose to give money to non-governmental wildlife conservation organisations when we see them taking action that we support.

In a politically devolved UK, wildlife conservation and most of the economic activities related to land use that affect wildlife – such as housing, agriculture, fisheries and forestry – are devolved matters. So when someone says that the government should do something about the biodiversity crisis they usually mean, whether they know it or not, the governments in Edinburgh, Cardiff and Belfast for those nations and the Westminster government for England and most international aspects of wildlife conservation.

Within each UK nation there is a statutory body with primary responsibility for wildlife conservation, whereas previously there was just one. The Nature Conservancy Council was disbanded in 1991 to be replaced by separate national bodies, and after over 30 years of further change we now have Natural England, Natural Resources Wales, NatureScot and the Northern Ireland Environment Agency, along with the Joint Nature Conservation Committee for a small number of UK-wide and international functions.

It is notable that whereas the NCC was almost exclusively a wildlife conservation agency that is no longer true of NE, NIEA, NRW or NatureScot, all of which have other responsibilities mixed in with, and often overwhelming, their wildlife roles. This weakens, probably deliberately, the statutory voice of wildlife conservation within the establishment machine. Trade-offs between human interests and wildlife interests must be made somewhere, but increasingly they are made within agencies with no democratic accountability and inside which the voice for wildlife is muted.

Outside the statutory sector there are a good many more familiar non-governmental organisations, mostly charities, that are clearly and specifically engaged in wildlife conservation. Here is a list of the main wildlife conservation organisations in the UK: A Rocha, Amphibian and Reptile Conservation, Badger Trust, Bat Conservation Trust, Buglife, Butterfly Conservation, Campaign for National Parks, Freshwater Habitats Trust (formerly Pondlife), John Muir Trust, Mammal Society, Marine Conservation Society, National Trust, National Trust for Scotland, People's Trust for Endangered Species, Plantlife, Rewilding Britain, Rivers Trust, Royal Society for the Protection of Birds, WildFish (formerly Salmon and Trout Conservation), Scottish Wild Beaver Trust, Shark Trust, Trees for Life, Whale and Dolphin Conservation, Wildfowl and Wetlands Trust, Wild Justice, Wildlife Trusts, Woodland Trust, World Wide Fund for Nature.

That's a somewhat controversial list as it does not include Greenpeace and Friends of the Earth (because to my mind they are little involved in wildlife conservation these days), or the League Against Cruel Sports (which is clearly and firmly in the animal welfare category), and nor does it include the British Trust for Ornithology (which is a science and monitoring organisation, and a very good one at that). More easily excluded is the British Association for Shooting and Conservation (from 1908 to 1981 the Wildfowlers' Association of Great Britain and Northern Ireland – note the interesting name change), as I can't see much in its activities that is aimed at conservation as defined in this chapter. But the Game and Wildlife Conservation Trust (from 1969 to 2007 the Game Conservancy Trust – note the clever name change) is a borderline case: much of its historical and some of its current work on lowland agriculture and waterways would merit the label of a conservation organisation but much of its recent behaviour and public stance, particularly in the uplands, would make it difficult to regard it as anything other than a force against wildlife conservation.

My list of wildlife conservation organisations is based on looking at the membership of the Wildlife and Countryside Link, Northern Ireland Environment Link, Scottish Environment Link, Wales Environment Link (none of which currently has BTO, BASC or GWCT as members). These umbrella organisations are liaison groups for like-minded organisations and do a lot of good work in agreeing policy positions across the range of organisations on a broad range of environmental, countryside and wildlife conservation issues. Have a look at the plethora of organisations involved and I'm sure you will come up with a slightly different list than mine, but however we choose to construct the border between NGOs which do and don't try to give wildlife a better future in the UK, we'll come down to a list of around 28 organisations to which the rest of us give our data, our time and our money in order to help them help wildlife at a population level. We do wildlife conservation vicariously, largely through our relationship with what I will call for brevity *The 28*, meaning the main 28 non-governmental wildlife conservation organisations operating in the UK.

Reflection 3

None of us is responsible for the state of the world at the moment when we are born into it, but we have to take some responsibility for its condition when we exit. I've spent most of my life working in the small, specialised area of wildlife conservation because of my fascination with wildlife and a wish to stop its steep decline. The wildlife losses I have witnessed have not been restricted to my country, nor to my lifetime, they are what we are doing everywhere and have been for ages. We did not arrive at the present by the most sustainable route, and on our journey we often destroyed wildlife for little benefit to people. I'm a conservationist because I want to help steer us to a better future.

Wildlife conservation aims to soften and ameliorate the net impact of all other human activities on the natural world. It is a technical business with jargon, detail and a need for biological information, but, alongside everything else it does, it is also about choices. These are not only the internal choices within the wildlife conservation world concerning what to do, and how to spend resources in the most effective way, but also, fundamentally, it is about what sort of world we want to bequeath to the future. I want that world to be richer in wildlife.

Caring about the environment and wildlife does not involve hating people. I like people, and I want them to have better lives. If my life had been otherwise then I might have ended up as a campaigner on social issues rather than environmental ones, for I want a future which encompasses greater equality and greater environmental riches together.

Looking at the world through the sustainable development lens, it's probably the case that we would all say we want to live in a world which is richer in material wealth, fairness and environmental assets (including wildlife). Can we have all of those things? The tricky bit comes when we have to make hard choices. How much material wealth would we sacrifice for a fairer society or a better environment? How much wildlife would we sacrifice for how much money? The temptation is to say that we can have it all, but history shows that economic development has won out and wildlife has lost out for a very long time, so something needs to change. Giving wildlife a voice at the decision-making table is what wildlife conservationists often find themselves doing, or at least trying to do.

Wildlife conservation is a social enterprise: we need to do it together if we are to make much difference. Whereas we can choose to enjoy the wildlife we encounter in our own homes and neighbour-hoods singly and without standing with anyone else, we can neither monitor the state of wildlife nor do much to affect it without standing with others who share our aims and ambitions. Being a naturalist can be a private pleasure, but being a wildlife conservationist means being a team player. We also need public support for wildlife conser-vation – not theoretical support but actual support, especially when difficult decisions need to be made.

We wildlife enthusiasts should realise that far from being a moderate and conservative agenda, wildlife conservation is a radical and progressive activity. To stand any chance of success, halting the loss of wildlife in the UK will require changes to the way that society operates, just as delivering a more equitable society for our fellow humans does. 'More wildlife' might seem like a modest ambition, just as 'more fairness' seems pretty uncontroversial, but whilst we may all sign up in principle to such values, and such a vision of the future, in practice it takes a huge amount of effort to engineer the necessary changes.

CHAPTER 4

Wildlife conservation successes

W e can learn from our mistakes but we can also learn from our successes. This chapter highlights 18 positive instances of wildlife conservation in the UK. It is deliberately an eclectic mix of big things and little things, quick things and slow things, and includes some examples which others might regard as failures but I prefer to regard as delayed successes.

It cannot possibly be a full account of all that has gone well in UK wildlife conservation, so each example is chosen to illustrate some general points – which are gathered together at the end of the chapter.

The 18 case studies are arranged alphabetically – so over the following pages you will be leaping from one subject to another in what I hope will be a stimulating manner. As you travel from policy to offshore island, and from species to habitat, just keep a note of which of the four horsemen of the ecological apocalypse are involved.

Agricultural policy

Farming covers around 70% of the UK land area and the wildlife on agricultural land is most generally in decline, but over the years we have put in place the means to make things better – even though this hasn't yet happened on anything like a large enough scale.

The changes in agricultural policy illuminate most of the characteristics of influencing public policy: it is slow to change, there are many vested interests involved, and yet over time real progress has accrued. Most people will roll their eyes at the thought of reading about the Common Agricultural Policy of the European Union, but that has been a major influence on farming and on the wildlife of the countryside for about 50 years so it can't be ignored lightly.

The birth of the CAP was at the very start of the European Economic Community. It was constructed by the six founder members (Belgium, France, Germany, Italy, Luxembourg and the Netherlands) back in the late 1950s, only a little more than a decade after its members had been at war with each other. The aim was to forge an economic partnership in which farming would be a part. It took until 1962 to put the CAP in place and it concentrated on food production, coming as it did after a period of food shortage and rationing. Security of food supply was its main aim, although even then policy makers had an eye on rural depopulation and social issues. Environmental protection was seen as separate from farming, and the EEC did not have a common environmental policy. The early CAP created a common internal market for farmed products, a common system of payment support to farmers, and import barriers to non-EEC farmed products. It is too simplistic to characterise the role of the CAP as a means of transferring German money to French farmers in return for other policy areas ensuring that German industry had access to French consumers, but that was indeed a significant element of the deal.

When the UK joined the EEC in 1973, UK farming came under the CAP. Farming had long been supported by subsidies in this country, but the nature of those subsidies changed. It became easier to sell our agricultural products within the EEC and for UK consumers to buy the farm products of our new economic partners (the original six members but also Ireland and Denmark), but it was more difficult to import products from our erstwhile close partners across the world (particularly Commonwealth countries) and to export to them.

The CAP was spectacularly successful in encouraging agricultural production. It achieved its goal through distorting the market by guaranteeing prices so that farmers didn't have to worry about finding a market for their goods. This led to a long period of over-production with wine lakes, and grain and butter mountains, where the taxpayer paid farmers for unwanted production. Land use was intensified, to the disbenefit of wildlife, because every grain of wheat and drop of milk had a guaranteed price and so it was worth farming more marginal land and investing in the means to do so whether it be drainage, larger machinery or the newest pesticides.

From the MacSharry reforms of 1992 to the present day (with a European Union of 27 states, until recently 28) the CAP has changed gradually to reduce agricultural surpluses, reduce payments to farmers, decouple payments from production and transfer more funds to environmental payments, mostly through voluntary schemes. We have seen the slow – though incomplete – greening of the CAP.

UK farming has become more intensive and more productive over the last 50 years and that isn't simply due to policy, it is due to a mix of policy and technological innovation. Mixed farming, where individual farms grow crops as well as raise livestock, is now much rarer, and the landscape has polarised geographically into the arable east and the livestock west. In arable areas, new herbicides, insecticides and fungicides have not only reduced losses of yield directly but have also allowed autumn, instead of spring, sowing of crops, and the stubble field has largely been lost from the land. Grass production is now fed by fertilisers, and silage production has become the norm. Hay meadows are an anachronism.

Now, in our post-Brexit existence, one of the greatest potential gains from leaving the EU is to set agriculture more quickly on a better, more sustainable, pathway. The byword is 'public money for public goods', and those public goods include twittering Linnets.

My local farmers can't sell the twittering of Linnets to me because I can listen to it from a public footpath and enjoy it for free. And indeed, if you stand next to me you can enjoy it too; my listening to that twittering doesn't deplete it for you. Linnet twittering is a non-market public good, and although it can't be bought or sold, it does have a real value; as do landscape beauty, carbon stored in the soil, and clean water running off the land. The past three decades of agriculture policy development in the EU have been spent in trying to make the CAP more friendly to these externalities, with limited success.

If we prioritise the population level of the Linnet in future agriculture policy then we will certainly get more Linnets – we may even get Linnet mountains as we formerly had wine lakes and grain mountains. But Lapwings need different things from the countryside than Linnets – so ought we to include them too? And every declining farmland bird? And declining farmland plants and insects? And wildlife is all very well, but what about carbon storage? And flood alleviation? And landscape values? And animal welfare? And land values? And good value for the taxpayer? Maybe we should shift support from agriculture to the NHS, or Trident missiles? And what about farmers? What impact on the farming industry will we tolerate in order to engineer different outcomes? And what value will we put on food security and home-grown food, or are we prepared to risk relying on imports from all around the world to feed the UK population?

These tricky decisions are thrown into sharper focus because we have left the EU. Until Brexit, UK governments have been part of the decision-making process for a large part of the European continent, but now we have to go it alone – that's what taking back control means. We do not yet have an entirely settled relationship with the EU but we are now living in a different world of imports and exports and a different policy agenda. It is difficult to imagine any government giving home food production a lower priority in those circumstances. We may find that the chlorinated chicken gets much more considera-tion in our future agriculture policy than the Linnet.

It's remarkable that in a generation or two we have turned around the way that public money pours into agriculture, from being a means to produce food with scant regard to the environmental impacts, to a means to support wildlife and the environment. Fifty-sixty years to stand agriculture policy on its head? That is about as quick as it ever could be – not as quick as wildlife needed, nor that the consumer needed, but far quicker than the agriculture industry would have wanted. UK farmland remains unfixed as yet, but the light at the end of the tunnel is burning brighter than ever before.

Ailsa Craig

The Brown Rat and the Black Rat are not native to the UK but both have been with us for a long time. Black Rats are all but extirpated from the UK but Brown Rats are almost everywhere, though the urban myth that you are never further away from a Brown Rat than 2 metres is just that, a myth. When I say that Brown Rats are everywhere, it isn't quite true. Just as has happened on remote islands around the world, rat eradication has been successful in the UK. This is clearly an area where welfarist and wildlife conservation values meet and clash head-on.

The island of Ailsa Craig in the Firth of Clyde is mostly famous for being a source of granite for curling stones but its conical hump, of over 300 m, is also a seabird colony with an impressive range of nesting species including Gannets, Razorbills, Guillemots, Black Guillemots and Puffins, Kittiwakes, Herring, Lesser Black-backed and Great Black-backed Gulls, Shags and Cormorants, and Fulmars. Around 100,000 seabirds nest just over 16 km off the Ayrshire coast.

It's a refuge for seabirds, away from people and surrounded by rich feeding grounds. And it's a long way for a Red Fox to swim, so you might expect that it is relatively predator-free – and so it would be, were it not for Brown Rats which were first spotted on the island when a lighthouse keeper's dog killed one on the jetty in 1889 while a coal ship was moored there. We forget, because none of us was there to remember, that Brown Rats only arrived in the UK in around 1728 and took quite a while to oust the also-non-native Black Rat from almost all of the country over the next century or so. Brown Rats only got to Scotland in appreciable numbers in the mid-nineteenth century, and then on to Ailsa Craig.

In October–December 1890 the lighthouse keepers killed over 900 Brown Rats, and they were found from the very top of the island down to the shore. The amount of seabird carrion produced in summer would keep a large rat population going into the winter months when the breeding seabirds had departed, but winter must have been tough for rats on Ailsa Craig.

If Brown Rats just wandered around eating berries, carrion and insects then there might not be any tale to tell, but they eat seabirds, mostly their eggs but chicks too. Puffins, living as they do in burrows easily reached by rats, rather than on cliffs, are very vulnerable, and a large Puffin colony declined from 'bewildering numbers ... so great that they darkened the sky' in the 1870s to 'practically extinct' by 1934, and the species was actually extirpated on the island soon after. And so things stayed until the early 1990s.

Studies of Ailsa Craig's nesting seabirds in the late 1980s and early 1990s showed that Fulmars and gulls had very low nesting success because of predation of the chicks by the Brown Rats, and nearby small islands had populations of Puffins, Storm Petrels and Manx Shearwaters, all of which would be expected to nest on Ailsa Craig (and were known to have done so in the more distant past).

In 1991, 3 tonnes of warfarin were air-lifted onto Ailsa Craig by RAF helicopters, and another 2 tonnes the next year. The poison was deployed in baits around the island in crevices and holes that would be used by rats and in specially constructed bait boxes so that care was taken not to lose baits to vulnerable non-target species. It worked – rats disappeared, ground-nesting species recolonised, cliff-nesting species extended their occupancy to areas that would have been vulnerable to rats, and in 2001 Puffins nested on Ailsa Craig for the first time in over 50 years. Puffins have since increased to over 100 pairs.

Animal welfare and wildlife conservation do not always walk down the same street hand in hand.

Flow Country

The Flow Country is in the very north of Scotland, in Caithness and Sutherland, almost touching the north coast. It is still a remote land of gently rolling moorland, much of it at a fairly low elevation covered in heather, Sphagnum moss and lochs and small pool systems. In May and June, this place is alive with breeding waders such as Golden Plover, Dunlin and Greenshank, with Red-throated and Black-throated Divers and Common Scoter nesting around the water bodies. For much of the rest of the year it is dark, cold and wet, and although there is wildlife to be found here, it doesn't shout at you like a displaying Greenshank will.

In the mid-1980s this was the scene of a dispute over land use. The fight was about the conservationist's friend, trees, but the trees in question were non-native conifers being planted across these boggy mires for profit. The key to making a profit here was not whether the trees would grow but a series of tax breaks and low land prices. The land was cheap because you couldn't do much with it except admire its ability to produce Golden Plover chicks at a better rate than most other places in the UK.

From 1915 until 1988 it was possible for any of us, if we owned some forestry, whatever rate of tax we paid, to choose whether our forestry enterprise was taxed under Schedule B (which exempted the sale of timber from tax) or under the normal business Schedule D (where the costs of establishing the forest could be claimed against tax). You had to choose whether you wanted jam, otherwise known as tax relief, today when you planted the trees or jam, otherwise known as tax relief, tomorrow – or more accurately in about 30–40 years' time – when you sold the timber. However, the tax system to operate under could be re-selected when the land changed hands – and that opened up the possibility of building it into a business model, pioneered by the Economic Forestry Group (the clue is in the name), where clients with different tax profiles were brought in at different stages so that the tax relief on the initial investment could be experienced by some and the tax relief on sales was enjoyed by others. This was legal and really quite clever, and thousands of hectares of previously pretty pristine deep-peat blanket bogs were ploughed, drained and planted over just a few years. The total area afforested in around a decade amounted to over 20,000 ha (that's 200 km^2, or an area 40 km long and 5 km wide).

The furore over the destruction of ancient out-of-the-way peatlands was big news, particularly because a number of high-profile celebrities – from the snooker player Alex 'Hurricane' Higgins to the TV and radio personality Terry Wogan – were involved as investors. And so the seeds of the end of the taxation system were set. In his Budget speech of March 1988 Chancellor Nigel Lawson not only put a penny on the cost of a pint of beer and strong cider, increased the fuel tax differential in favour of unleaded petrol, reduced the standard rate and all the existing higher rates of income tax, but also abolished the tax relief on forestry establishment costs and essentially made

Schedule B the norm by exempting timber sales from taxation. He commented that:

> the present system cannot be justified. It enables
> top rate taxpayers in particular to shelter other
> income from tax, by setting it against expenditure
> on forestry, while the proceeds from any eventual
> sale are almost tax free. The time has come to
> bring it to an end. I propose to do so by the simple
> expedient of taking commercial woodlands out of
> the income tax system altogether ... expenditure on
> commercial woodlands will no longer be allowed
> as a deduction for income tax and corporation tax.
> But, equally, receipts from the sale of trees or felled
> timber will no longer be liable to tax.

The change in the forestry tax regime was a big win for wildlife conservation and was seen as the power of the conservation case winning out over commercial interests. And so it was. However, the tax system had been in place since 1915 and nobody had ploughed up the Flow Country to any great extent until the early 1980s, and certainly not on the deep peats that were richest for wildlife. No, the impetus for afforestation here was a combination of many things. In a crowded island like the UK many of the places where it was possible to plant trees, and where the owners of the land wanted to plant trees, already had trees on them, so attention was being focused on the more and more marginal areas, such as the Flow Country. And the Flow Country was a large area, some of it owned by rich people who themselves could benefit from the existing tax regimes, and the scale of the overall potential land-holdings made the deployment of large, expensive machinery more feasible than if the scale of opportunity had been smaller. And land prices were low. All these things conspired to prompt imaginative entrepreneurial firms, in this case mostly Fountain Forestry, to take a leap of imagination and strike while the opportunity existed.

If you look at a map of wildlife conservation designations for the Flow Country you will see that it is pretty much plastered with Sites of Special Scientific Interest, Special Protection Areas and Special Areas of Conservation, but there are some strangely shaped gaps in the maps, some with rather straight boundaries. Closer inspection will reveal that these are the areas of forest that were planted in the 1980s, and you might wonder why those areas were left unprotected and so allowed forestry in. But if you do wonder that you've put the cart before the horse, because all too often the forests came first and the protected areas came after and had to fit in with the forests. The Sletill Peatlands, Rumsdale Peatlands, Strathmore Peatlands and Ben Griams SSSIs all were either designated or greatly extended in the early 1990s as part of the aftermath of the stramash over the Flow Country. If only they had been there before, and if only they had covered much of the 20,000 ha of land that disappeared under trees in those frantic years of the late 1980s.

The rapid afforestation of the Flow Country, and its potential destruction as a wilderness area (with huge carbon stores) and a very important site for wildlife, is often seen as an unintended consequence of a tax system and a forestry grant system which was quickly changed. However, it was also, and perhaps primarily, a consequence of incomplete notification of protected areas for wildlife conserva-tion. If the area had been covered with the range of SSSIs and SACs and SPAs that now cover it and its blanket bogs, no forestry company would have dived in to make a quick buck from the system. I'm sure that the process of upland site designation was carried out according to an apparently sensible set of priorities, but it was the conservation-ists who accidentally let in the foresters, not the Treasury. Are our most important sites for wildlife conservation adequately designated and notified today, I wonder? Where are the gaps in the system now?

If the battle of the Flows had occurred nowadays it would never have been a battle – the carbon stores would have won the day without a Greenshank needing to be mentioned.

Fonseca's Seed Fly

There isn't an awful lot to say about Fonseca's Seed Fly, but there ought to be, and that is a commentary on the state of wildlife conservation in the UK. I hadn't even heard of it until I asked some mates for examples to include in this book. This small brown-grey fly lives only in the UK, only in Scotland and only, as far as we know, along an 8 km stretch of coast on the Dornoch Firth in Sutherland. So, it is a UK endemic species with a tiny world range, and in 2018 it was added by the International Union for the Conservation of Nature (IUCN) to the list of globally threatened species.

A large part of this species' world range was threatened by a potential development – not a supermarket or a new road, but a golf course in the sand dunes of Coul Links between Loch Fleet to the north and Dornoch to the south. Highland Council foolishly consented the golf course development, but after the proposal was called in by the Scottish government that decision was overruled. I think that was the right decision and it was a triumph for how the planning system should work, and sometimes does. It wasn't just the Fonseca's Seed Fly that persuaded the decision makers in the end, for the site also has rare plants, such as Purple Milk-vetch and Coralroot Orchid, and a coastal Juniper grove as well as quite decent birds. A coalition of wildlife organisations (RSPB, Buglife, Plantlife, Butterfly Conservation, Marine Conservation Society, National Trust for Scotland and the Scottish Wildlife Trust) mounted a campaign to press home the views of those species and the battle was won. Strikingly, there are simply loads of other golf courses in the area and this is a site that has several layers of protection for its wildlife conservation importance. The councillors took the apparently craven option of approving the development proposal against the recommendation of their own staff and of the Scottish statutory wildlife conservation advisor (then Scottish Natural Heritage, now rebranded as NatureScot) perhaps in the knowledge that central government wouldn't let it go ahead but none of the blame would stick to them.

We can wonder how many sites might just have endemic, or at the very least extremely rare, species but their presence is not known or recognised by 'the system'. This won't often happen with birds these days, or probably plants, but it must be a real risk with invertebrates. You can't protect what you don't know is there. But on the other hand, maybe with more investment in the right survey work we'd be finding this little fly all over the place – which would be good for the species but a handicap to using it to protect all the other wildlife on Coul Links. Either way, this clearly calls for more investment in invertebrate survey and monitoring.

This was a triumph of NGOs working together – that's a good thing. And it was a small triumph, in the end, for the planning system, although Highland Council wasted loads of other people's time by not doing the right and obvious, and legally required, thing straight off. We can also file this away as a classic case where the developer thought they could make money and hoped that they could, almost literally (but not literally) bulldoze their way through the planning system through force of effort. I don't blame the developers too much, it's what I expect them to do, driven by the profit motive. Their view might be 'It's up to the planning system to decide, but let's give it a roll of the dice. After all, Donald Trump got permission to build a golf course on the east coast of Scotland under not wholly different circumstances.' Let's not expect too much social responsibility from commercial interests; for many of them the three-legged stool of sustainable development is actually a shooting stick of profit.

Hedgerows

Hedgerows are a very familiar part of the British lowland countryside and they are very British (and Irish). You don't see so many hedgerows as you travel through the rest of Europe. They are a product of our history of land use. Hedges are the remains, in some cases ancient remains, of the woodland that once covered far more of the UK. Hedges are linear scrublands and woodlands which form natural(-ish) corridors along which wildlife can travel through what can be an inhospitable countryside.

But hedges don't exist as a gift to wildlife, they have an agricultural and social purpose. In many cases hedges are there to be stock-proof barriers, which calls for thick spiky hedges that can keep in (or out) the most cunning and determined of horses, cattle or sheep. On livestock farms, hedges are therefore an asset as well as a wildlife resource. In arable areas, hedges are less useful as they take up space which could grow a crop, may harbour wildlife that the farmer regards as weeds or pests, though on the positive side they may release useful predators of insect pests into the crop for free.

Changes in agriculture have influenced how many hedges we can now see in the countryside. As the mechanisation of agriculture increased in the 1960s the hedge became less and less welcome in arable areas, many of them in the east of the country. If you are spraying or harvesting your wheat with a big machine you need big fields, ideally with straight edges in order to work most efficiently. A network of small fields of strange shapes is not ideal, so, in the absence of much protection, hedgerows were removed by farmers to increase the profitability of their businesses. Maybe half of Britain's hedges were lost in the 60 years after the Second World War.

My Breeding Bird Survey (BBS) survey route includes a boundary hedge that separates Northamptonshire and Cambridgeshire, and it is the bit of Cambridgeshire that was once Huntingdonshire, which is the old county often mentioned as the one which suffered the highest hedgerow loss. I've known it for the past 25 years and most of the birds I record are living mostly in the hedgerows.

Public uproar at the changing face of the countryside resulted in better protection of hedges and led to the Hedgerow Regulations Act of 1999, which requires landowners, essentially farmers (as garden hedges are exempt), in England and Wales to get permission before they remove any hedgerow. It's now quite difficult to remove a hedge – your best chance is if it's a short, new hedgerow with little wildlife; otherwise you will probably have to live with it. That actually sounds quite draconian to me, that in a changing world of agriculture you are now stuck with the field boundaries that your parents or grandparents could live with and not necessarily the ones you want now. But there doesn't seem to be that much uproar from the farmers about it. Putting in a new hedge is a lot easier and you may well get a grant for that. And once you have a hedgerow then there is loads of free advice on how best to manage it, some guidelines on the same subject, and also regulations about what you mustn't do.

It's not a perfect situation, but if we are looking for a compromise then this might possibly be it. A planned compromise along the lines of 'conserve the very best hedges but you can get rid of any of the worst that you like' might have delivered more wildlife, but the actual 'oh crikey, half of our hedges have gone so we'd better clamp down' approach is much better than nothing.

Hope Farm and Skylark patches

Hope Farm is the slightly cringe-making name for a bog-standard arable farm that the RSPB has owned and managed (through contractors) for over 20 years. The aim of the project was to manage ordinary farmland so that its wildlife would increase and yet the farming would remain profitable and thus act as a model for other farmers facing similar issues, with the added bonus that maybe those in charge of agricultural policy would use the results to inform grants and regulations.

Buying a farm in the realm of the barley barons near Cambridge, even a fairly small one, along with its farmhouse and buildings, cost a seven-figure sum so it wasn't a trivial investment of money, effort or reputation. However, the declines in farmland birds had been researched and by the turn of the millennium were pretty well understood as far as arable farming systems were concerned. There was a good chance that bird populations could be dramatically increased without any similarly dramatic loss of yield or profitability. And that is just what happened: the number of farmland birds breeding on Hope Farm increased threefold. Let's just pause there – threefold, in 10 years, and then those numbers were maintained over the next decade. That's no flash in the pan. Looking closer at the species that have contributed most to this reveals that Skylark, Linnet and Yellowhammer have all increased a lot, which tells any birder that both the fields themselves and the surrounding hedgerows must be in good nick, and that both invertebrates and seeds must be in good supply. Winter bird numbers have increased spectacularly too, with a more than fivefold increase in numbers.

Several management practices helped the increase, but one of the most interesting, and simplest, was the creation of Skylark patches. Research had shown that Skylark nesting success in cereal fields was very poor – they simply weren't producing enough young each year because they weren't nesting often enough. Skylarks regularly used to nest three times during a summer season, with successful nests being scattered across that period. In modern cereal fields, they did well on their first nesting attempt, when the crop was young and not too tall or thick, but through the season they were pushed to nesting along the tramlines in the crop (the rows where machinery tyres habitually travel), where they were lost to predators that also travelled along those rows and killed by the machinery itself. Creating small non-crop patches within the wheat provided nesting refuges and enhanced feeding locations for Skylarks. This very simple method worked very well, used a tiny proportion of the cropped areas, and increased nesting success and productivity enough for Skylark numbers to soar at Hope Farm.

But what of the yields and profits (they aren't the same, but are fairly closely linked)? They did well too. The wildlife gain has not been bought at the price of a great loss of income, and despite the vicissitudes of weather, pests and changes in government grants and policies, Hope Farm remains in the average of profitability for local farms despite taking about 15% of its land out of production and producing loads of breeding and wintering birds. If the same were done across all UK arable farmland, the farmland bird index would be higher than at any time in our lifetimes and farming profitability hardly affected.

This is a great example of how the environmental leg of sustainability can be lengthened without cutting off anything much from the other two legs. Of course a farmer can squeeze a little more profit out of the land by ignoring the wildlife impacts – that's what has been happening for years. But this is a choice, and not such a stark choice as it might first appear. Farmers are making the decision whether to remove wildlife from their farms and make a little more money, or to make a little less money and keep much more wildlife.

Hope Farm is a local success but it could have been, and could be in future, replicated on a much larger scale with very impressive results through building it into agricultural best practice. Does that make it a success or a failure? Let's call it an overly delayed success.

Knepp

The 1,400 ha Knepp estate in West Sussex is a well-known success story of do-it-yourself successful rewilding by private landowners. Charlie Burrell and Isabella Tree had a failing farm which couldn't really make ends meet, and with the benefit of large injections of public money, as well as through getting expert advice, they turned it into what is essentially a wildlife reserve through letting wildlife have its way. It's a remarkable story and a great success on two fronts – the wildlife and the economics.

The wildlife dividend has been spectacular, with rare species such as Nightingales and Purple Emperors thriving, and fast-declining species such as Turtle Doves showing very positive population trajectories. This has been achieved through ceasing arable farming, managing extensive grazing and restoring a more natural water regime.

On the back of this burst of wildlife and an interesting and different landscape, Knepp is now a visitor attraction and tourist destination. The variety of accommodation options include yurts, shepherds' huts, bell tents and treehouses. There is meat for sale which is 'free-roaming, pasture-fed, organic Old English longhorn cattle, Tamworth pigs, and red and fallow deer'. And you can visit on a series of safaris to see the Purple Emperors, listen to Nightingales, see the rutting deer and learn more about rewilding in general and at Knepp in particular.

Knepp has hosted a somewhat controversial reintroduction project of White Storks and plans to bring Beavers and Red-backed Shrikes into the mix soon. One can imagine a series of exciting new wildlife initiatives as the site develops – and they may indeed be necessary to maintain the visitor flow.

Knepp is an entrepreneurial one-off. The ecological lessons from 20 years' rewilding are profound. Few would have expected the explosion of wildlife that has resulted through, basically, scrubbing up the land. The ecological benefits could probably be replicated in a large number of other sites across southern England. However, the economic model could rarely be replicated as it is based on a considerable input of public funds and a unique visitor experience. You can't just stroll into Knepp and listen to the Nightingales; you must go on a Nightingale safari – you are pretty much guaranteed to hear them, but you will be charged £95 a head (although dinner – I'm told it's delicious – is included). People will go home after a visit to Knepp talking about the wonderful meat, the amazing hut/yurt/treehouse they stayed in and enthusing about the wildlife in a way that they are unlikely to do after a visit to most wildlife reserves. To the extent that the Knepp model is a financial success (who knows, without seeing the books?) it is based on wildlife success accompanied by very successful marketing. The marketing is brilliant on the website, in the book and in the media. Look at Knepp as the Longleat of wildlife reserves and you are getting close to the model. And that's the main reason why this is not a great financial model for future projects – Knepp took the considerable business risks and now, having made a success of things, has the first-mover advantage. Yes, it can be followed in other localities but it will be difficult for any other estate within, say, 150 km of Knepp to do something very similar because they will be in competition with Knepp.

How do we get more Knepps? The barriers are not ecological, they are simply economic, because few landowners would be able to make money out of such an enterprise. I doubt that there is much room for many private-enterprise Knepps, but there is loads of scope for more ecological Knepp-like success stories. They would have to be set up under a different financial model. One possibility would be for wildlife charities to buy land to manage in this way, but another would be for the state to acquire land for these purposes. A score of Knepps would cost the equivalent of the loose change that you might find down the back of the UK's sofa, or that of any of the constituent nations. The cost would be modest – so what's stopping our public bodies from doing it?

Lead ammunition

Lead is a useful metal because it is fairly non-reactive and is malleable and so can be worked into shapes. But it is a poison, and so we have largely removed its use from paints, water pipes, fishing weights and petrol. A last stronghold of lead use is the shooting industry, which has fought moves to persuade it to switch to non-toxic ammunition.

As far back as 1983 the Royal Commission on Environmental Pollution recommended in a report that lead should be removed from fuels and paints and that, basically, we should reduce any and all uses of lead on environmental and human health grounds. Much of that has been done, with lead being removed from fuels in the UK in 1999 after a period when all new cars had to have catalytic converters to allow the use of lead-free fuel. Anglers have got used to using other weights in almost all of their fishing since the 1980s, and bodies such as Forestry England and Forestry and Land Scotland have almost totally banished the use of lead ammunition in culling deer in their forests (the venison from which goes into the human food chain).

The use of lead shot for recreational wildfowling – shooting ducks, geese and waders in wetland habitats – was banned in slightly different ways in each of the four UK nations in the late 1990s and early 2000s, but surveys of ammunition used to kill wildfowl on sale in game dealers and butchers show that compliance with the regulations is very poor, only approaching 50%. In any case, the restriction does not apply to the major sources of spent ammunition in the countryside, namely shooting Pheasants, Red-legged Partridges, Red Grouse, Grey Partridges, and so-called pest species such as Woodpigeons.

Since the RCEP report the evidence for harm to people and wildlife from lead has strengthened further, and the Labour government in 2010 set up a panel to look at the use of lead ammunition again. That expert group recommended in its 2015 report to government, now a Conservative government, that use of lead ammunition should be phased out – as had already happened in some other countries and in some states of the USA. On her last day in the job as Secretary of State at Defra, and while David Cameron was making his farewell speech after the Brexit referendum result, Liz Truss rejected the report on entirely spurious grounds. Despite government inaction, there are signs that some retailers are moving to label game meat as potentially containing lead, a poison, on their shelves and more importantly are requiring their game suppliers to use the readily available non-toxic alternatives to lead – most notably steel, bismuth and antimony.

I've been involved in a moderate way in the campaign to remove lead ammunition from our environment and our food for the last 40 years and I am at a bit of a loss to understand why it is taking so long. There are many people who eat game meat, probably more than you realise, and it would be a good idea if that number grew – but what reputable organisation would promote the increased consumption of a meat which has impacts on human health and whilst the shooting of it spreads a toxic metal far and wide? The shooting industry has opposed the move towards a switch to non-toxic alternatives with a determination and a venom which seems out of all proportion to the slight inconveniences it would cause them (temporary slightly higher costs and the need to adjust shooting techniques). They prefer to shoot toxic metals into food and the environment.

We have governments to referee such debates and to come down on the side of the science and the public good. It's notable that the UK moved pretty quickly to get rid of lead from petrol – a much more important, but a much more difficult transition for industry and the public than to get lead out of ammunition which is shot into the environment and into our food. The pressure on decision makers to act on lead in game meat and as an environmental pollutant is mounting all the time, and I expect almost all use of lead ammunition to be a thing of the past by the mid-2020s.

Marine Protected Areas

The marine environment is a neglected world for wildlife conserva-tion. When we stand on the seashore, we don't see that much wildlife because we can't see beneath the surface, we don't see the damage caused by extractive industries and fisheries, and we don't measure the changes in wildlife through our own lifetime experiences.

A helpful difference between the marine environment and the terrestrial one is that wildlife reserves in the sea can benefit the fisheries which are excluded from using them. A marine reserve that excludes fishing effort can act as a secure breeding ground which returns commercially valuable wildlife to the sea outside the reserve. That's the theory, and it has been shown to work many times across the world and a few times in the UK. It's very different from farming on land, where crops and livestock are not wildlife, and so each area of land excluded from farming represents a loss of farming income.

A marine wildlife reserve adjacent to Lundy Island at the mouth of the Bristol Channel was the UK's first statutory no-take zone (NTZ). This was essentially marine rewilding. A major human intervention to the site, fishing, was removed and wildlife could recover from un-restricted exploitation and from habitat damage caused by fishing operations. Comparing the densities and sizes of Common Lobsters in the reserve with those in adjacent areas that still had some fishing showed that lobsters in the NTZ were larger and more numerous. Common Lobsters were also thought to be spilling out of the NTZ and repopulating areas where the fishery persisted, thus leading to a more sustainable long-term fishery.

The experience in the waters around Lundy and elsewhere in the world generally makes the case for Marine Protected Areas, Highly Protected Marine Areas or marine wildlife reserves – however you want to describe them. Towards the end of the last Labour UK government, in 2009, the Marine and Coastal Access Act was passed, with support from the opposition Conservative Party. The Act aimed to establish coastal rights of way and also made 'provision in relation to marine functions and activities' – which sounds rather bland, but those provisions included the establishment of a new body called the Marine Management Organisation (MMO) to administer activities in England's seas, and envisaged the setting up of a network of marine wildlife reserves to protect wildlife.

The Act is by no means a cracking read (I certainly don't recommend reading it for enjoyment), but it was an important step forward for marine wildlife. By 2020, the MMO was able to trumpet that there were 175 Marine Protected Areas (MPAs) around just England's seas and coasts, and in the same year a review chaired by former Defra minister Richard Benyon promoted the idea of Highly Protected Marine Areas to add more effectiveness to the network. The network of 175 sites is a mixed bag and includes some designations which were already in existence and some that are coastal rather than marine but, let's be fair, they all count as steps forward. The Marine Conservation

Zones (MCZs) were rolled out with the first 27 at the end of 2013, another 23 in 2016 and a further 41 (making 91 in all) in 2019. Each tranche of sites added around 10,000 km² to the area of MCZs. The total of 30,000 km² is equivalent to the combined area of England's four largest counties (North Yorkshire, Lincolnshire, Cumbria and Devon), but the area of sea under the ambit of the MMO (inshore English waters and offshore UK waters) is 230,000 km².

Taking a step back, there has been considerable progress in the last 15 years to give the marine environment a greater measure of the type of protection that is so important, and rather taken for granted on land. This is real progress that passes most of us by because we are not 50 km offshore very often, and even if we were, the sea wouldn't look any different – no fences, no notices and no patrolling staff, but now with significantly greater protection.

No Mow May

Cutting the grass is such an ingrained habit that many a sunny summer Sunday morning is marked by the sound of mowers mowing in my street – but this is a habit of the gardener subjugating wildlife rather than the naturalist loving wildlife. Yet that is changing, as even the most assiduously mown plot of land may now have an unmown strip for several weeks in late spring and early summer. A new habit, quite trendy in fact, of not-mowing has grown up, and people now boast of the buttercups that are to be found in those patches and the insects to be heard buzzing there as a pre-lunch drink is taken outdoors.

No Mow May was the brain child of Dr Trevor Dines of Plantlife, and the simple model of leaving a garden lawn unmown for the month of May really seems to work well both for people and for wildlife. The campaign started in 2019 and has grown each year. The rationale is that easing off the mowing allows nectar-producing plants to flourish in a way that they will not if decapitated every Sunday morning in a quest for the 'perfect' manicured, striped patch of very short grass. Anything that requires less work and which is good for wildlife is a winner, and the results are immediate and obvious to participants. First, you can see that you have made a difference, your garden looks different and then you'll find yourself looking at the flowers that emerge from the new regime and the insects that visit them with pride and interest. Lockdown, in the spring and summer of 2020, was ideal for getting people to focus on the place that was their own a lot more, and many people used lockdown to ensure not that they had a perfectly manicured lawn but that they had more wildlife in their gardens. Walking along the street, past houses whose owners we didn't know, we could see patches, strips or whole lawns left unmown that had always had the normal treatment in the past, and so No Mow May grows as a noticeable badge of wildlife action – you don't need to boast about what you are doing (though some do), for everyone can see for themselves.

Plantlife say that participants across the country reported a total of 200 species of plant growing in their lawns and that the nectar produced amounted to the equivalent of food for 400 bees per lawn per day for the majority of lawns. And some 20% of participants had superlawns that were feeding the equivalent of 4,000 bees per day. All those Daisies, Dandelions and other plants supplied their nectar as garden owners basked in the sun and in the rosy glow of feeling that they had made a difference for wildlife.

The challenge now is to maintain the momentum and spread the word to more individuals – but also to businesses and local authorities. Why shouldn't a piece of every publicly owned grassland in the country, at least in some years, benefit from a similar regime?

At a time when rewilding is all the rage, and the loss of insects from our countryside is becoming obvious to many, putting them together in a labour-saving personal action is genius – particularly because it works. No Mow May provides almost instant proof of success to the participants. Your patch of temporarily unmown grass is noticed by your friends, your neighbours, people walking their dogs, delivery drivers and many more. If anything powerfully embodies the idea that every little helps, then this is it.

Otters

Otters disappeared from many of our rivers in the mid-1950s and early 1960s, largely, it seems, because of the impacts of organochlorine cyclodiene pesticides such as aldrin and dieldrin in arable farmland. Mammals, being sneaky nocturnal creatures, can decrease in numbers without it being very obvious to most of us – we rarely see them anyway so it is hard to tell if we are seeing them less. But after a while the experts and those most predisposed to notice them (such as anglers) did sound the alarm. Otters did not decline in mainly pastoral areas such as Devon and much of central and north Wales, and their populations remained strong on the island of Ireland – the areas of decline were very much coincident with arable farming.

Compared with the rate of loss, the return was slow, strangely slow considering that use of the relevant cyclodiene organic insecticides was banned in the 1960s. There were few signs of recovery by the late 1970s at the time of the first national (England) Otter survey, when only 6% of 2,094 sites which were known to have had Otters in the past seemed still to have them. Further national surveys showed increases to 10% (1984–86), 23% (1991–94), 36% (2000–02) and 59% (2009–10). This strongly suggests that habitat quality was not good enough over large areas of England for a decent recovery to be mounted. Levels of toxic heavy metals such as copper, lead, cadmium and mercury took much longer to fall, with much of the progress not coming until after the 1990s. It may also be that although cyclodienes were very probably important in the early years, the rise of polychlorinated biphenyls (PCBs) became a more important deleterious factor for Otters after the 1960s and slowed the rate of recovery. A range of other factors including disturbance and low fish stocks are also likely to have played a part. The relative contributions of different factors to the decline are still not entirely resolved.

As Otters returned they found themselves more cherished and loved than they had been before. Two films, *Ring of Bright Water* (1969) and *Tarka the Otter* (1979, but based on a 1927 book) had brought Otters to people in a way that nothing else could, and the Otter Trust (founded 1971) also played an important role in turning the Otter into a much-loved and much-welcomed creature. Otters received full legal protection (from, for example, hunting by dogs) in 1978 and their holts were protected under the Wildlife and Countryside Act of 1981. Finally, in 2015, Kent reported the return of the Otter – and that made a full set of English counties recolonised from the west and the north. Now Otters can cavort in waterways in our towns in a way that they never did before.

There is the possibility, though it does not seem to be very well supported by hard facts, that as Otters return the numbers of non-native American Mink are reduced – and if they are then that is good news for the greatly depleted numbers of Water Voles.

Peregrines and pesticides

The Peregrine Falcon did just what a species should do if it wants to draw attention to its decline – its population fell off a cliff. Suddenly, in the 1950s, Peregrines in North America and Europe were producing very few young and the populations were in freefall. As a canary in the coal mine it seemed to be doing a great job. The cause of the decline took quite a few years to prove, and then it took even longer to forge a solution through international effort, but it has worked. The Peregrine Falcon is more numerous now than it has been in living memory – which isn't an insignificant thing to be able to say of a species whose global extinction was feared. When I was born, the possibility of me having a conversation with the local garage staff about the Peregrine that often perches on our local church spire would have been inconceivable.

In fact, in the UK, it was mass die-offs of widespread farmland birds in agricultural situations that first triggered alarm, and a Committee on Toxic Chemicals was set up which established that the problem was probably the use of organochlorine cyclodiene chemicals such as aldrin, dieldrin and heptochlor. Those chemicals, which are not naturally occurring, were used as insecticidal seed dressings and they got into farmland bird populations mainly through ingestion of the treated seeds. After that, a wide range of birds of prey, not just Peregrines but Sparrowhawks and others – and Otters too – were being poisoned by chemicals in their prey. Dichlorodiphenyltrichloroethane (DDT), another organochlorine chemical, was also fingered for some of the blame as it was shown to cause eggshell thinning in raptors such as Peregrines and Sparrowhawks – the eggshells were often so thin that the incubating female would simply break her own eggs when trying to incubate them. In Europe it seems that the decline was caused by cyclodienes assisted by DDT, whereas in North America, because of differences in chemical use in agriculture, it was probably DDT assisted by cyclodienes.

Poisoned birds lying in the fields is not a good look for an industry that grows food for human consumption, but despite that, the chemical industry fought restrictions on the use of their chemicals tooth, and nail and for many years. The end result was pretty inevitable, however, and use of those chemicals is now banned in many but not quite all countries in the world. A near-global ban on the use of a chemical isn't a very common event but it happened in this case. The chemicals in question aren't good for human health either, and high levels will cause death, but the human impacts have been slow to emerge; it was the impacts on wildlife that really led to their withdrawal from use.

Testing of chemicals before their widespread use has been pretty lax really, and it's not very reassuring to realise that testing is never going to be perfect in any case. Yes, we can test things and, yes, we ought to have more stringent testing in some cases, but we'll never test for everything that might happen – recent examples include neonicotinoid pesticides, hydrofluorocarbons and the ozone layer, veterinary drugs and Asian vulture declines and the continuing widespread use of lead ammunition (see above). It is a measure of the diversity and complexity of wildlife that different bits of it will react to challenges in different ways. That's not at all surprising. What is encouraging is that we haven't been awful at reacting to problems that slip through the net. Yes, we need a better net, and yes, we need to react better to emergencies, but overall we could have done far worse.

Pine Martens

The Pine Marten is a member of the weasel family which lives in woods and feeds on a wide range of vertebrate prey but also berries, and visits gardens to feed on nuts where it has also been found to have a particular fondness for peanut butter). It occurred widely across the UK mainland and on some of the larger islands until the rise of gamekeeping and sporting estates in the mid-nineteenth century caused a rapid population decline and loss of range. Deforestation would have added to its problems. It is now mostly found in well-forested upland areas in Scotland, northern England, Wales and some parts of Northern Ireland.

Pine Martens are now spreading naturally, and aided by releases, into their former range. They occur in many forested areas of continental Europe and you should expect to see them in a wood near you some time in the next few decades. The Pine Marten is a kind of mammalian Red Kite, a predator persecuted to near extinction over 100 years, and unfamiliar to people living in places where it once was common, but now making a comeback, partly naturally and partly aided by us. Like the kite, it is a beautiful creature, but the largely nocturnal habits of mammals means that the return of the Pine Marten will go unnoticed by the casual observer far longer than did the resurgence of the Red Kite.

Reintroduction projects seem to be effective: 51 Pine Martens were released into forests in mid-Wales between 2015 and 2017 and they appear now to be well established and breeding successfully; 35 were released into the Forest of Dean, Gloucestershire, in 2019–21 and successful breeding has occurred. A natural range extension is also occurring: Scottish Pine Martens are crossing the border into Kielder Forest and the Lake District.

Such an increase in range is good news simply because it represents the return of a beautiful native mammal, but there are two remarkable extra reasons to welcome the Pine Marten's resurgence.

First, as Pine Martens increase in numbers the non-native Grey Squirrel decreases. This has been shown in studies in Ireland, Scotland and the north of England. It's a dramatic impact too, not a slight fall in Grey Squirrel numbers but a massive reduction, which pleases foresters, who regard Grey Squirrels as pests. When Grey Squirrels were released into the British countryside in the late nineteenth and early twentieth centuries they encountered a world largely lacking in Pine Martens, and there is no other widespread native predator which will create a landscape of fear for Grey Squirrels like the Pine

Marten – although Goshawk (another persecuted predator making a comeback) would certainly add to their nervousness. Grey Squirrels, being heavier and less agile than Red Squirrels, often move from tree to tree via the ground, and that's when Pine Martens can pounce. Grey Squirrel dreys are so obvious and accessible that young Greys must be vulnerable to predation as well. We should imagine Pine Martens killing quite a lot of Grey Squirrels but also putting the wind up them so that they alter their behaviour to avoid predation risk. The return of the Pine Marten means the removal of Grey Squirrels – which is probably the cheapest form of pest control known to humankind.

There is more! As Grey Squirrels decline in abundance, Red Squirrels move back to occupy the squirrel niche. This, too, is well documented. So we get three gains for the price of one, or the price of none where Pine Martens are increasing on their own without reintroductions. It is true that Pine Martens eat Red Squirrels, but we can already see that they live in balance over much of their shared UK and European continental ranges, and Pine Martens and Red Squirrels also coexist in recolonised areas.

It's a remarkable story. All the shooting of Grey Squirrels has done nothing like as much good as a few boxes of Pine Martens, and all the releases of Red Squirrels won't do as much good for Red Squirrels unless there are some boxes of Pine Martens released at around the same time, but ideally a few years before. I don't recall anyone trumpeting this as a likely outcome in the past. The law of unexpected consequences runs both ways – it can deliver nice surprises as well as nasty ones. The Pine Marten ought to be one of the poster-boys of ecological restoration and rewilding.

Protection of Birds Act 1954

The Protection of Birds Act 1954 (PBA) was a progressive and ambitious piece of legislation that came into being during a Conservative government led by Winston Churchill. Its provisions remain in force largely unchanged nearly 70 years later despite us joining and leaving the EU during that period and despite devolution of such matters to the four legislative bodies of the UK in the meantime. The PBA formed some of the basis for the 1981 Wildlife and Countryside Act (WCA), which took on board protection needed for all taxa, not just birds, and the implications of the UK's joining of the European Economic Community in 1973, and the provisions of the Birds Directive in 1979.

The PBA is unusual in being a product of the Private Members' Bill system, which is essentially a ballot of backbench MPs in which those whose names are drawn near the top of the list get the opportunity to introduce legislation into parliament – and in most cases the proposed Bill comes to naught. In fourth place in the ballot on this occasion was Priscilla, Lady Tweedsmuir, the Unionist and then Conservative MP for Aberdeen South (1946–66) and subsequently Baroness Tweedsmuir in the House of Lords, one of fewer than 30 female MPs at the time and only the 54th woman MP to be elected. A prominent supporter of the Bill was Captain Tufton Beamish MP (Conservative, Lewes, 1945–74), who was still in parliament, as Baron Chelwood, to support the Wildlife and Countryside Act of 1981.

Lady Tweedsmuir introduced her Bill despite the fact that there was already a government Bill aiming to do similar things introduced in the House of Lords. Both Bills aimed to regularise the wide range of existing legislation which protected wild birds, their eggs and nests, from deliberate harm such as shooting, trapping and egg collecting for food or acquisitiveness. At the time there were 26 separate Wild Bird Protection Acts and over 250 regulations in force which covered different geographic areas, specified different open and close seasons for quarry species, and protected the eggs of different species to different extents in adjacent counties. As Tufton Beamish said in debate, the Avocet made a good choice when it started its recolonisation of England in Suffolk, where it was listed as protected, rather than in adjacent Cambridgeshire, where it could legally have been shot. Some of the replaced legislation dated back to the eighteenth-century reign of George III.

Both Bills aimed to clear up the confusion of what was and wasn't protected, and when, by establishing a simple UK-wide system of bird protection. The starting point was that all wild birds and their nests and eggs were protected by law, with rare species having even stricter protection, and with two relatively small groups of birds having lesser protection. Those exceptions were of gamebirds, which the Game Acts going back into several previous centuries had defined, which could be shot for fun under certain conditions within their own open seasons (post-breeding season and for varying times into the winter) and a small group of species variously called vermin, pests or nuisance birds. Previous Wild Bird Protection Acts had tended to list those species that were protected, which put species at the mercy of the breadth and depth of legislators' ornithological knowledge and of changes to bird distributions over time. The PBA, by protecting all species as its starting point, was essentially future-proofed to new species.

Having lists of species with greater protection, gamebirds and nuisance birds allows, maybe even encourages, tinkering with the lists over time, and that should be regarded as a good thing, if done well, since much has changed in wildlife terms in the almost seven decades since the PBA came into force. Over that period, the PBA and its later manifestation in the WCA, has stood the test of a long period of time. It was remarked upon at the time, and is even more striking now, that the wording of the Bill was clear, concise and comprehensible. Unlike many modern legislative tangles, it is a model of legislative drafting.

Where Lady Tweedsmuir's went further than the government Bill (which was withdrawn as her Bill gained much greater support) was in protecting all nests and eggs from collecting. The government proposed that common birds should be exempted from such protection on the grounds that they didn't need it and that egg collecting was a harmless pursuit which encouraged an interest in natural history, but the Bill's supporters argued for simplicity and clarity and also that egg collecting was a thing not to be encouraged in any fashion. The PBA turned an incredibly common boyhood (vastly more than girlhood) hobby into a criminal offence overnight.

I recommend reading the passages of Hansard where the Bill was debated – they bring to mind a long-gone age, but it was the forethought of one of the early women MPs, along with the support of several of her male colleagues, which established the system of bird protection that we still have today. Part of the success of the Bill was probably that different interest groups – ornithologists, conservationists and wildfowlers – came to an agreement on what was needed. The essence of those distant deliberations lives on in all parts of the UK today. Successive governments could have done away with it completely, or made radical changes, but they didn't – at least not yet.

Reform or abolition of driven grouse shooting

Grouse shooting is recreational shooting of native, wild, Red Grouse on the heather moors of northern England and eastern and southern Scotland, and a very few hills in Wales, Northern Ireland and elsewhere in Scotland. About 500,000 Red Grouse are shot in an average year, some of them are eaten, and individuals will pay thousands of pounds each day for the experience of shooting Red Grouse, so there is a small economic benefit to grouse shooting, which may feel big to some small local communities but which doesn't add up to a hill of beans in broader economic terms.

The Red Grouse are wild birds, not farmed, but producing unnaturally high densities for shooting demands very intensive management of species and the habitat on a landscape scale. Their natural predators are relentlessly controlled (because a Red Fox won't pay £85 for each Red Grouse she despatches whereas a shooter will), the heather on which the grouse depend for food and cover is managed through burning and draining of moorland areas, and the birds are dosed with medicines added to grit that they ingest to aid their digestion of the tough heather shoots. This management acci-dentally favours a few other species, such as some wading birds, but it has a wide variety of ecological disbenefits too. The scale of predator control is huge and not only targets native predators which the law allows you to kill but very often embraces illegality with the killing of Badgers, Hedgehogs, Red Kites, falcons, eagles and harriers. Intensive burning of blanket bog, a protected habitat, not only damages the habitat but also makes it more prone to dry out and the peat below the surface to emit greenhouse gases. The Committee on Climate Change has called for a cessation of vegetation burning on peat soils, just the types of places where grouse shooting takes place, simply on greenhouse gas emission considerations – and that really ought to be an end of the discussion right there.

Burning the vegetation for the benefit of shooters also reduces a moorland's ability to store water and increases the likelihood and scale of flooding downstream. Having your home or business flooded because water pours off the grouse moors is not just an economic impact, it is a social harm. It is literally a rich person's hobby having an impact on other people's lives – that is a social, or anti-social, impact.

For all these reasons, and more, wildlife conservationists, climate campaigners and animal welfare campaigners have all promoted an end to driven grouse shooting. In the last nine years, and due to many people's efforts, progress has been made. The Scottish government has committed to licensing both the business of grouse shooting and the burning of vegetation, and in England Defra has brought in measures, feeble ones, to limit vegetation burning. English local authorities (e.g. the City of Bradford Metropolitan District Council on Ilkley Moor) and some businesses (e.g. Yorkshire Water) have moved away from allowing intensive grouse shooting on their land.

A 'traditional' hobby of a very small number of people, which occupies a surprisingly large amount of land, much of it in our National Parks, is under the cosh. Assailed by those who don't like recreational shooting, those who do like birds of prey, those who don't like their homes being flooded, those who favour rewilding of the uplands, and those who want carbon stores to be fully protected, the reform of intensive grouse shooting is well under way. And in England, progress has been made even under a Conservative government.

This is, under any sort of rational assessment, remarkably rapid progress against a powerful vested interest of wealthy well-connected landowners. It's an example of change that has further to go, but has come a very long way in a short time. What was, not long ago, regarded as a quaint hobby for the rich or landed is now seen as an issue – a wildlife issue, an environmental issue, a social issue, an economic issue and a political issue. As an issue it won't go away, but as a land use I am sure that it will, and fairly soon.

And when driven grouse shooting ceases, we won't miss it. The economy will not be affected, rich people will spend their money on other hobbies, flood damage will be lessened, insurance costs will decline, harmful carbon emissions will be reduced, and life will go on.

Sites of Special Scientific Interest

Sites of Special Scientific Interest (SSSIs) are sites which are designated (notified is the technically correct word) as special places for wildlife by the state, and that designation confers some protection on those sites. Their name – SSSIs in England, Wales and Scotland, Areas of Special Scientific Interest (ASSI) in Northern Ireland – is probably a big turn-off. SSSIs came into existence thanks to the National Parks and Access to the Countryside Act 1949 and were simply called 'wildlife reserves'. They were fiddled about with in the Wildlife and Countryside Act 1981 and again in the Countryside and Rights of Way 2000 (CROW) Act but the essence of their role has remained more or less constant over more than 70 years. The aim was to select some of the best sites of particular habitat types in each area of the UK so as to protect just some places from damage. They were never meant to protect all the good places, although it is a testament to their effectiveness that many of the sites that might have qualified back in the 1950s, but weren't protected, are now lost under the plough or built development, with the result that the network today covers a much higher proportion of important sites than it once did – a strange outcome to celebrate. After devolution the notification of these sites and their protection became a responsibility of each of the four UK administrations. As well as wildlife sites there are also geological sites in the network.

It is a biological reality that some places have more wildlife than others and some species are more clumped or restricted in distribution than others. SSSIs and other forms of site protection are particularly appropriate for such places and species. The few locations of rare plants or insects, the best of our seabird colonies, estuaries and reservoirs teeming with migratory birds, and the very best of much diminished habitats are good candidates for being SSSIs. They are easily delineated and a relatively uncontroversial line can be drawn on a map to indicate where activities injurious to the wildlife conservation interest of the site should be limited.

Roughly speaking, about 6,500 SSSIs cover around 10% of the UK land area and confer some protection on the wildlife within their boundaries. Most of these sites are still owned by private landowners, who feel a mixture of pride and prejudice about the designation, and for each there is a list of dos and a much longer list of don'ts associated with the site's management.

When SSSIs were first envisaged in the post-war era, they were seen as prime examples of our cultural heritage and as places where wildlife could be studied by boffins. Their role in research and study was mentioned ahead of their role in protecting flora, fauna, geology and geomorphology. Later, as pressures on the countryside grew, they became more important in preventing fine wildlife sites being lost completely – woods cut down, grasslands ploughed up, wetlands drained – but in time it became clear that there needed to be a way not just of preventing bad things from happening, but of encouraging maintenance of good management. The activities that had allowed wildlife to persist on a site into the 1950s might not be very attractive to landowners in the late 1990s, even if they were the grandchildren of the original landowners. The CROW Act of 2000 introduced measures to ensure good management and prevent damaging management (including neglect) of such sites. In the first years of this millennium, after the passing of the 2000 Act, there was much activity from major statutory and non-statutory landowners to ensure that SSSIs were in favourable ecological condition. Targets were set, standards were set, monitoring was done and meetings were held. Bodies such as the Forestry Commission, the National Trust and the Ministry of Defence vied with each other to do a good job and to show the way for the many private landowners who were perhaps less motivated to put their all into site management, even if grants were provided.

Two pieces of EU legislation, the Birds Directive of 1979 (which in some ways was a supercharged version of the Protection of Birds Act

1954) and the Habitats and Species Directive of 1992 (which in some ways was a supercharged version of the Wildlife and Countryside Act 1981) required the UK to designate protected areas for wildlife, and that necessitated a very significant expansion of the existing network of wildlife sites. The SSSIs first established under the 1949 act formed the basis of the sites under the Birds Directive (Special Protection Areas, SPAs) and under the Habitats Directive (Special Areas of Conservation, SACs). In the UK all SPAs and SACs are also SSSIs, but not all SSSIs by any means (only about 50% by area) are SPAs or SACs. And just to round that off, many SPAs and SACs have essentially the same boundaries.

The 2010 Lawton review of the adequacy of wildlife sites to cope with a changing climate looked at statutory wildlife designations as well as wildlife reserves but overall concluded that (for England, but it will apply across the UK) we need bigger, better and more joined-up wildlife sites. This is just what the NGO landowning community has been working on for a century or so.

Since the passing of the 2000 CROW Act (which applies to England), annual figures have been published on the proportion of SSSIs in Favourable condition (the definition of which varies with habitats and sites and is complex, but realistically complex). In 2003 44% of English SSSIs were deemed to be in Favourable condition, but by 2021 this had fallen to 38%. Fewer than half of sites, notified by government as being of wildlife conservation importance, are in good condition. Imagine if that applied to schools or hospitals! On the face of it, the good news is that in that same period the proportion of sites that are in Unfavourable condition but recovering has increased from 13% in 2003 to 53% in 2021, which sounds pretty good. But what counts as recovering can sometimes be real but sometimes merely amounts to there being a plan in existence. Even this measure has fallen from a peak in 2011.

The idea of protecting the best sites for wildlife conservation, enshrined in the 1949 Act, lives on and forms the basis for site protection to protect wildlife across the UK more than 70 years later. Without the SSSI system, significantly updated and improved in 1981 and 2000, our wildlife would be far worse off. It strikes me as remarkable that a system dreamt up when clothes rationing ended in the UK remains in place, and is still doing good. Although there is the possibility that laws can be changed and weakened, history shows – both in our enterprise of wildlife conservation and elsewhere – that generally they are strengthened and improved.

Stone-curlews

The Stone-curlew, or Goggle-eyed Plover, is neither a curlew nor a plover but it is a wading bird with large staring eyes that tends to live in stony places. It's a bird of farmland, but only some types of farmland. Away from the UK you mostly find Stone-curlews in dry plains of Mediterranean countries and the steppes of eastern Europe – places with big skies hanging over extensive grasslands, often of a dry and stony nature. And the Stone-curlew is found in such places in the UK too: the sandy Breckland of East Anglia, the grassy Salisbury Plain and the dry heathlands of coastal East Anglia and Dorset, Hampshire and Sussex. The Stone-curlew's UK range matches closely with a geological map of sandy or chalky soils and so it's always been a bird of restricted range, but characteristic of dry grasslands. The large eyes are because it feeds a lot at night, snapping up invertebrates, and its curlew name is because its call, most often noticeable at night when all else is quiet, resembles that of the Curlew.

A couple of centuries ago there were maybe a few thousand pairs of Stone-curlews in the UK, but by the mid-twentieth century the numbers were around 1,000+ pairs, which fell to some 150 pairs by the 1980s. The causes of the decline were loss of the grass and heather heaths that the birds favoured but also increasing loss of eggs and chicks through agricultural work in the fields in which they nested. The habitat loss was a failure of the system to protect enough heathland, valuable for a great many species, in the face of agricultural intensification, the planting of trees by the relatively newly established Forestry Commission and the insidious creep of built development as our population increased. Similar declines, for similar reasons, occurred elsewhere in the bird's European range. On farmland, the losses of chicks and eggs to agricultural machinery were unintended consequences of routine farming operations; Stone-curlew eggs and chicks are very well camouflaged.

A recovery project, led by the RSPB and involving scores of cooperative farmers, allowed nests to be found and marked by volunteers, and this resulted in far fewer young birds being lost. Since the 1980s the Stone-curlew population has doubled to over 300 pairs after decades of decline. Now farming grants are available to create fallow plots within fields, where conditions are made good for Stone-curlews so that they are attracted to nesting in these safe locations and farmers are compensated for the loss of income and the work involved. These are essentially nest boxes for Stone-curlews, but the fallow plots also benefit other declining farmland birds, such as Lapwings and Skylarks, as well as rare arable plants.

Stopping a long-term decline in numbers and doubling the UK population in 30 years is a significant achievement, and one that few conservation organisations are able to claim. The Stone-curlew sits alongside the Cirl Bunting (a formerly widespread farmland bird now more or less restricted to southwest England) and the Corncrake (a grassland bird that used to inhabit every UK county but is now largely restricted to those parts of northwest Scotland where hay is still harvested in traditional ways) as examples of interventions which have turned around the fortunes of farmland birds in spectacularly successful ways. But all these species are still in intensive care – if the effort flags, or the money is reduced, then they will resume their declines and could be lost very rapidly. Their recovery still depends on ongoing intervention – in a way that the recovery of Pine Martens and Peregrines no longer does.

Wildlife reserves

Land ownership gives you power of management, and that's what wildlife reserves are – patches of land owned or managed by bodies (or sometimes individuals) with the aim of protecting or improving wildlife in that area. The four main land-owning and land-managing wildlife conservation organisations in the UK are the National Trust, the RSPB, the Wildlife Trusts and the National Trust for Scotland, which were, respectively, founded in 1895, 1889, 1912 and 1931. From a standing start, in around a century they have amassed land-holdings of over 500,000 ha or around 2% of the UK land mass. The National Trust manages 250,000 ha of land in England, Wales and Northern Ireland, the RSPB 130,000 ha across all four UK countries, the Wildlife Trusts 100,000 ha across the UK, and the NTS 73,000 ha in Scotland. From its inception in 1983 the John Muir Trust now has a land-holding of 25,000 ha, which is a very significant area too. Other wildlife conservation NGOs also have smaller, though often important, land-holdings. Some of this area is leased rather than owned, but in very few cases are those leases short-term or highly caveated and so they amount to control of the management of the land.

Land ownership gives you the ability to manage land according to your own principles, means and objectives, but it doesn't give complete control over the ecological horsemen of the apocalypse. Habitat loss and overexploitation of wildlife are pretty well controlled by land ownership. Nobody can build a housing estate on your land without you noticing or agreeing, and land ownership usually brings the power to prevent hunting, shooting and fishing – but wildlife reserves can be considered ecological islands set in seas of various levels of choppiness and danger. That housing estate may arrive next to your wildlife reserve and there won't be a huge amount that you can do about it. Pollution, including climate change, will wash over your wildlife reserve as everywhere else when the wind blows and the rain falls, and the likes of Japanese Knotweed and Grey Squirrels won't turn tail at the boundary of a reserve either. The wood where I used to hear Nightingales is managed by a wildlife conservation organisation but the Nightingales no longer fill it with song.

Are our wildlife reserves doing a good job? They most certainly are. One very easy way to demonstrate that is by reviewing where you go when you want to see some wildlife. Much of my wildlife viewing, locally and nationally, is done on wildlife reserves owned by conservation organisations. That's partly because these are welcoming places, sometimes with car parks, toilets and refreshments, and because many are set up to display the wildlife they have to good effect, with paths and hides and information provided – but it is primarily because they are rich in wildlife.

These places exist largely because of civil society not because of the state – the area of land managed as wildlife reserves by statutory agencies is very small. It would be difficult to conceive of the state of UK wildlife without all the wildlife reserves owned and managed by charitable conservation organisations. If they all disappeared tomorrow our wildlife would be greatly further impoverished. And yet they occupy only around 2% of our land area – their positive influence is out of all proportion to the paucity of their coverage. How would we feel if each of them was on average twice as big? I think we'd love it.

If you want to see and experience wildlife then increasingly you will spend your time on wildlife reserves, those relatively tiny areas that provide effective shelter from the threats to wildlife. These sites show, in an analogous way to Herb-Robert popping up in the crack between the pavement and my wall, that wildlife doesn't need that much help to thrive, it just needs a chance.

Reflection 4

Wildlife conservation can be a frustrating activity, but it is far from being a hopeless one. We know what to do to give wildlife what it needs, and we have helped many species on a local scale, increased the populations of a few species on a national scale, and protected many of the best places for wildlife in the UK.

We've been busy. We've reintroduced species, restored habitats, campaigned successfully for better wildlife protection laws, prevented the loss of wildlife hotspots and intervened to reduce the harmful impacts of land uses such as forestry, fishing and farming on wildlife. Wildlife conservation has matured and passed through the proof-of-concept stage, and all that needs to happen now, the only thing, is that it needs to be rolled out on a big enough scale to allow wildlife to thrive around us in a way that would delight and surprise us.

The foregoing accounts exemplify the nature of our victories over the four horsemen of the ecological apocalypse – some were quite big victories, others brief skirmishes. Non-native species were central to the Ailsa Craig story but also featured in surprising ways in events involving Pine Martens and Otters. Overharvesting was part of the justification for the introduction of the Protection of Birds Act all those decades ago, and more recently for Marine Protected Areas, and to some extent for SSSIs and Wildlife Reserves, as well as being a part of the justification for reforming driven grouse shooting – where illegal 'overharvesting' of wildlife causes declines of protected species. Pollution, whether of lead or by agricultural chemicals, seems to be an issue which takes a long time to reverse. However, most of the examples here, from No Mow May and Fonseca's Seed Fly to agricultural policy and the Flow Country, involve protecting, improving or recreating good habitat for wildlife. That's probably roughly representative of the distribution of conservation effort.

When I was born, wildlife conservation largely consisted of protecting wildlife from people who wanted to harm it. Wildlife conservationists guarded the sites of rare birds, butterflies or plants from collectors and set up small wildlife reserves. Now we still do these things, on a much bigger scale, but we also attempt to tackle major economic activities, not only through seeking to halt or amend individual development schemes (such as roads, housing, ports or golf courses) but also through building wildlife protection into all aspects of our lives from food production to energy production. Whereas wildlife conservation used to be small-scale and impinged on few people's lives, it is now attempting to be much bigger in scale, which means that it aims to and will affect many more people's lives. Wildlife conservationists used to be seen as the underdogs, combatting bad people doing bad things, but now there is much more scope for them, us, me, to be seen as meddling in everything and being a bit of a pain. This view is most likely to be promulgated by vested interests who want their own financial ambitions to trump wildlife needs, but it can be shared by a wider range of more neutral individuals.

I won't go through them all, but each of the conservation success stories in this chapter was opposed by some body or some interest group. The Pine Marten, a native species that has been missing from much of our countryside for many decades, has been portrayed as a vicious predator (their teeth do look sharp, I agree), which will do untold damage to our wildlife if encouraged to return, by those who fear for the fate of their captive-bred and released non-native gamebirds. Those who wish to make money out of yet another new golf course argue that destroying wildlife-rich habitat will bring healthy exercise and tourist revenue to the local area. Farmers may claim that any restraint on their activities which benefits wildlife will make them poor and increase starvation in the world. But if we look coolly at the inconvenient impacts of wildlife conservation on our lives as a whole, they are trivial.

We could have five times as many wildlife reserves as we currently do and there would be no discernible harmful impact on the national economy. If we had hundreds more Hope Farms, and scores more Knepps, then UK agricultural production would feel a pin prick not a pain. The return of the Peregrine Falcon, Otter and Pine Marten to our towns, rivers and forests has not reduced the size of the UK economy, and we can be sure that the return of White-tailed Eagles, Beavers and Lynx wouldn't either.

The restoration of Red Kites to our skies can be taken as a model for conservation progress. When wildlife is missing, few regret its absence, but when it returns it is widely celebrated and appreciated. The transition from absence to presence is opposed by some, often quite vehemently, but when wildlife recovers we soon wonder why we didn't get it back quicker.

So why is wildlife conservation so difficult to achieve at a larger scale? That is the subject of the next chapter.

CHAPTER 5

Why are we failing so badly?

I describe wildlife conservation as a social enterprise because I think such a description captures an oft-ignored element: the fact that none of us can do it successfully alone but together we can. The threats to wildlife come from our collective action and inaction, and a renaissance of UK wildlife will only come from our collective positive action.

A pertinent question in looking forward to a better deal for wildlife is to ask whether the existing means of harnessing collective action, most notably government action and the action of wildlife conservation organisations, are fit for the job in hand. Surely, the lesson of decades of continuing losses of wildlife in the UK must be that our recent approach to wildlife conservation has not met the need, and it's a pretty strong signal that we must do better in future. Are we doing the wrong things, doing the right things badly, or just not doing the right things often enough?

Why might we be failing?

There's a long list of potential reasons why we, as a society, are failing to protect and enhance UK wildlife. They are familiar reasons because they resemble the list of excuses that we use ourselves for the failure to achieve much more everyday tasks. They are of the following types: 'I didn't realise there was a problem'; 'I didn't know how to fix it'; 'I didn't have the power to fix it'; 'I didn't have the time to fix it'; 'I thought someone else was fixing it'; 'I tried to fix it, but it didn't work'; 'I'm not really that bothered about fixing it'.

Clearly there is no single solution that can be applied across all land and sea areas to restore more of our lost wildlife. Skylarks need different solutions than do blanket bogs, and Pine Martens need different solutions from Fonseca's Seed Flies. But this is a good thing – we can get on with fixing the things we know how to fix without having to wait for global solutions. And to an extent that is what has happened, as the examples in the previous chapter illustrate. And yet wildlife is still declining in the UK.

Lack of knowledge is not the problem. We know that wildlife is declining. The information is better for some habitats and some groups of species than for others, but the fact that wildlife is in decline is not in dispute. And we know the general reasons for this – they are the four horsemen of the ecological apocalypse playing across habitats and species in different combinations and measures. For each site or species, the combination may be subtly different, but not so different that we are clueless what to do. And doing something that seems sensible is often the best way forward, to see whether or not it works. Be somewhat sceptical of the scientist who says that 'more research is needed', and point out that it might be better simply to invest in more certain solutions in different places and for different species rather than to invest in another three-year research project.

I think that we, collectively, as a society – so I am primarily referring to government and its agencies here – have failed in two main areas. The first is through woefully inadequate investment in solutions, and the second is through the deployment, too often, of weak mechanisms for change rather than strong measures which would be much more effective. We could be in a much better place than we are, and I blame government for that.

Our wildlife NGOs bear some of the responsibility too, but a much, much smaller share than government because their power to affect change on the ground is so much less. However, a fragmented sector which is easily distracted, and which fails to deploy the power of its membership to best advantage, must take some of the blame for the current parlous situation.

And to the extent that we are all voters and many of us are members of those wildlife charities, then we all share tiny fractions of responsibility too. And I, as a former employee of a wildlife conservation organisation, share a little bit more than most of the rest of you.

Lack of government investment

Governments do wildlife conservation, and government action is essential to success. Wildlife conservation is seen as a 'good thing' by politicians in the sense that wildlife is so well loved by the population, particularly after a brilliant Attenborough or Packham series on TV, that governments want to appear to be doing something positive on the wildlife conservation part of the environmental agenda. However, if you look closely, they aren't very active.

Look at the various government departments on and around Whitehall. If you put them through a sustainable development sieve, then the vast majority of them are there to deal with the economic and social aspects of sustainable development of one species, our own. Very roughly, economic development is lodged in the roles of the Chancellor of the Exchequer, the International Trade Secretary, bits of the Foreign Secretary, bits of the Food and Rural Affairs Secretary, the Communities and Local Government Secretary, and the Business and Industrial Strategy Secretary, while social development resides with the Home Secretary, Education Secretary, Health and Social Services Secretary, Justice Secretary, bits of the Food and Rural Affairs Secretary, Defence Secretary, Transport Secretary, Work and Pensions Secretary, Digital, Culture, Media and Sport Secretary, and Minister for Women and Equalities. So where does environment reside? It sits inside a small part of the Department for Business, Energy and Industrial Strategy and a small part of the Department for Environment, Food and Rural Affairs.

The architecture of government shows scant regard for the environment, and that bit of environmental concern that is wildlife conservation is almost completely hidden from view. It cuts no ice at all to suggest that the environment, and wildlife conservation, are integral parts of the work of all government departments when, by casual and detailed observation, they quite clearly are not.

Just to rub it in, if you consider that the UK governments together spent £842 billion in 2019/20 (that's £842,000,000,000) spread over 11 categories (one of which is 'other') then the top two, Social Protection and Health, account for half of the spend, and Education,

Debt Interest, Defence, and Public Order and Safety account for another quarter. In tenth place (out of 11, one of which is 'other') comes not Environment, but Housing and Environment (<4% of the total government spend), and somewhere in there must be the paltry investment in the environmental aspects of sustainable development, of which a small part is wildlife conservation. There's no doubt that spending on environment is low on the list of priorities according to government spending figures. It amounts to maybe 2% of total government spending. Is that how you expected the spend to be arranged – or had you, like me, not thought that much about it before?

The Defra budget for 2020/21 was £8 billion, but nowhere near all of that annual budget is spent on wildlife conservation – it will be a small and arguable fraction. There will be other environmental expenditure in other Whitehall departments, and more in the budgets of the devolved governments. But only small fractions of this small fraction are spent on wildlife and its recovery.

The structure of government makes little space for environmental matters and the budgetary allocation even less, and the politicians with responsibility rarely bring passion and even more rarely useful knowledge to the decision-making table. This would be less serious if only governments used their full powers in the right way.

The unused powers of governments

We are losing wildlife at crisis levels in the UK because too many bad things happen and not enough good things. Government has the power and the resources to rectify this but only if it acts as a player, not a bystander, and that is a stance that is unpopular on ideological grounds in Westminster.

Lack of effective regulation. Much of our lives is regulated: paying taxes is not voluntary, speed limits are not voluntary, various antisocial activities such as fraud and murder are not simply discouraged but are illegal. All this is familiar, and probably seems reasonable to most of us. Yet in many environmental areas we exhort or incentivise people to do the right thing when we should simply tell them what to do.

The small state is a political dogma ill suited to halting a wildlife crisis, or a climate crisis, and there is plenty of evidence that it fails. One of the most interesting NGO reports of the last decade was by RSPB economists who examined the effectiveness of voluntary approaches in delivering change across a wide range of issues, not just environmental, and across many countries. Voluntary approaches could fail in three main ways: lack of ambition, lack of delivery when adopted, and lack of adoption by enough people. And if a voluntary initiative fails in any such way then it is likely to fail overall, and the review showed that was the case with many such initiatives. It's unsurprising that the voluntary approach is so loved by vested interests: a three-year voluntary initiative is likely to postpone any regulation by a good four years or longer. Far from being a step forward it is often marking time or falling backwards. Recent failed voluntary initiatives in the UK include a switch away from peat compost, cessation of burning of blanket bogs, and phasing out lead ammunition.

Laws which are not enforced aren't much use. The laws that protect birds of prey from poisoning, trapping or shooting are palpably not properly enforced in the UK today, and so illegal activity by shooting interests – illegal since before I was born – carries on at a scale that limits the geographic range and population levels of species such as Golden Eagle, Hen Harrier, Goshawk, Peregrine Falcon, Buzzard and others. Similarly, there is lax enforcement of regulations about pesticide use, pollution of waterways and the use of lead ammunition. In practice, criminals act against environmental interests and get away with far too much.

Lack of land ownership. One of the easiest, most tried and tested and most neglected forms of action is to buy chunks of the landscape and manage them primarily for wildlife, but in the UK the state neglects to deliver public benefits through such a straightforward and proven delivery mechanism.

Public land ownership in the UK is very low and has become lower in recent decades as the state has divested itself of land. The successor bodies of the former UK Forestry Commission are the largest public owners of land in the UK but the Ministry of Defence has large areas too. However, public land ownership is probably only

10% in the UK (it's difficult to tell exactly); after all, 70% or more of the UK is farmland and the vast majority of that is privately owned, and another 10% or so has been developed for transport, industry and residential property, with 13% as woodland (some of which is publicly owned). In comparison, federal lands in the USA account for almost a third of the total land area and are concentrated in the west of the country and Alaska where deserts and mountains predominate – and, generally, wildlife thrives.

Government policy in the UK seems to be based on the idea of influencing landowners as a major tool in wildlife conservation – and yet it doesn't use regulation enough, and hardly enforces the regulations that do exist in many places where they would benefit wildlife. Meanwhile, government does nothing to increase its land ownership, which would allow it to become a much more significant player in its own right. It's almost as if government isn't really trying.

Vested interests

Vested interests are ones where the individual or organisation involved in influencing a decision has something more to gain, usually financially, from some possible outcomes than from others.

The main beneficiaries of planting trees all over the Flow Country were high-rate taxpayers and the companies that ran the forestry businesses, but of course the arguments deployed were all about the value to local communities. The default position when dealing with a developer, an industry or a business should be to challenge every word, every supposed fact and every projection for all you are worth.

Farming is an example of a vested interest which is a prime target for reform if one seeks a better deal for wildlife. It is an ultra-conservative industry floating on a sea of public subsidy and fighting hard to retain that subsidy for as long as possible with as few strings as possible.

Farming produces commodities such as wheat, meat and oils for the food industry – the main customer of the agriculture. Farmers, are paid by the tonne, litre or whatever for their commodities and so the incentive is to maximise production, provided that the costs of production don't eat into the profits. That's fair enough, but no farmers are paid by the food industry for the number of Corn Buntings, Corncockles or Harvest Mice on their land. Nor is the farmer paid for the amount of carbon stored or the purity of the water that runs off the land. The market is blind to these matters and does not take a sustainable development approach.

In discussion with government, farming does not act like a think tank looking for the best environmental outcome or the best value for money for the taxpayer. It acts like most other money-grabbing industries and deploys a familiar range of arguments about farmers being poor, farming feeding the world, that farming subsidies create cheap food, and that UK farmers are the best in the world. The proponents of such arguments are the farming unions, of which the National Farmers' Union in England and Wales is the most politically adept and has major influence on farming policy.

Over the past few years, agricultural interests have persuaded government not to replace set-aside with environmental schemes, have watered down the effectiveness of wildlife grant schemes, have promoted landscape-scale Badger killing, have asked for let-offs over bans on the use of harmful pesticides, and have minimised actions to combat climate change – and in all these cases farming interests have been opponents of wildlife conservation. We would have much more wildlife if the farming unions were weak and the environmental groups were strong. The power of farming to perpetuate its own financial interests is massive, and it comes not through individual farmers making the case but from collective action.

Farming is an effective vested interest which perpetuates and increases the losses of wildlife, but similar arguments apply to forestry, the shooting industry, the house-building industry, the pesticide industry, the aviation industry, the fishing industry, the fossil fuel industry and others. Just look at the 2021 COP26 meeting in Glasgow, where the fossil fuel industry had more lobbyists than did the whole of civil society. Do we believe that those lobbyists were making the case for the public good?

Vested interests are everywhere, are very powerful, and are not overly scrupulous in arguing their case. We can't really blame them for that, but we won't talk them round either.

Wildlife conservation is not seen as a political issue

I wrote a monthly column entitled 'The political birder' for 10 years for the excellent magazine *Birdwatch*. I was often told that wildlife conservation was not a political issue and even more often that it shouldn't be a political subject. Neither is right, and the first view is simply factually incorrect to an astounding degree – but the widespread currency of these views is a real block to making conservation progress.

Politics is all about how the world should be, and what are the appropriate ways to make it so – there's an awful lot of that in wildlife conservation. You may think that all political parties should be keen on conserving wildlife, but even if we could get to a position where that were true then different political parties would favour different routes to the same goal. One of the biggest schisms in politics is over the role of the state. I am a confirmed lefty and believe we need a wise, fair and powerful state to make decisions that are for the public good. I'm not the person best able to tell you what those on the right think (because as far as I'm concerned, it is nonsense), but it involves reducing the size of the state, trusting markets and the good will of individuals, and the UK forging ahead on its own because we are so great. Even without my lapse of impartiality, it is clear that there are radically different visions of how best to run the country, and it's not surprising that this extends to different views on the best ways of conserving wildlife and solving environmental issues. These views are hard-wired into people's belief systems. So, yes, wildlife conservation is a political issue, and yes, it is a party political issue, because one political party may have better solutions than another.

One of the ways to judge between political parties and their potential impact on wildlife is to listen to what they say and read their election manifestos – hardly anyone does that, though!

Let me give you an example from after the 2010 general election. When giving a talk, I would often ask the audience whether they had ever read the manifesto of a political party, and very few hands went up. I'd then ask whether they approved of the Badger cull which began to be rolled out by the coalition government in 2013 and which had been off the table while Labour was in power. There was hardly anyone who approved of the Badger cull in a room full of naturalists. Then I pointed out that the Conservative manifesto had included the words 'As part of a package of measures, we will introduce a carefully managed and science-led policy of badger control in areas with high and persistent levels of bTB.' Those words very clearly signalled a move to mass killing. 'Science-led', 'carefully managed' and 'a package of measures' have all largely gone by the board but the writing was on the wall, or more precisely on page 97 of the 118-page Conservative manifesto published in April 2010 entitled *Invitation to join the government of Britain*. The Labour manifesto in 2010 was almost completely silent on environmental matters aside from climate change, but their position was well established as against culling but in favour of vaccination. Few people would have voted at the fag end of a Labour government solely on the fate of Badgers, but no-one who voted can blame the parties for not giving them a clear choice. That's usually the way with environmental and wildlife issues, even if you need to read between the lines.

By 2025 over a quarter of a million Badgers will have been culled in England, which would not have been culled if there had been a different government in charge. Some Badger enthusiasts voted unknowingly for that cull because they didn't think that wildlife conservation was a political issue. Badger culling was strongly promoted by the farming industry, who persuaded Conservative politicians to take that route. Wildlife conservationists did not win this battle, and it was a political battle.

Whenever I have turned, hopefully, to the manifesto pages that might set out the aims of a future government for the 70,000 UK species of plant and animal then I have found, sandwiched between quite a lot on climate change (these days) and a substantial amount on the welfare of wild and domestic animals, a few slightly stilted sentences on wildlife conservation. That's been my experience for decades. It's as if, and it's precisely as if, politicians think that the voters are vaguely interested in wildlife and its conservation but that they leave all those thoughts behind when they cast their votes. Could they be right?

Governments have massive power over our wildlife, and that same power can be deployed for massive harm or massive good.

Conservationists, meanwhile, don't have a grip on that power, and without it our wildlife will continue to decline.

A house divided – a case study

Let us turn to our wildlife conservation organisations. Are they the solution to the chronic wildlife decline? Let's limber up with a case study to explore some of these issues – the Wildlife Trusts.

The Wildlife Trusts are one of the Big 3 of *The 28*, but actually they are 47 independent UK entities: there are two national Wildlife Trusts (for Scotland and Northern Ireland), some regional Wildlife Trusts (such as Bedfordshire, Cambridgeshire and Northamptonshire; South and West Wales; Hertfordshire and Middlesex), many county Wildlife Trusts (such as Norfolk, Northumberland and Nottingham-shire) and some smaller ones such as Isles of Scilly Wildlife Trust and Sheffield Wildlife Trust (as well as the non-UK Manx Wildlife Trust and Alderney Wildlife Trust). It's an interesting take on conservation geography, isn't it?

In theory, the rationale behind this proliferation of bodies is that individual Wildlife Trusts are in tune with local people and local issues but they can also raise their multiple voices together on issues that affect any of the four UK nations or the UK as a whole. The umbrella body, the Royal Society of Wildlife Trusts (generally known simply as the Wildlife Trusts), is a kind of secretariat, spokesperson and leader for everyone else. It's a system that works to some extent. There is always a tension, perhaps a productive tension, between the folk in Newark and everyone else – and it is a more difficult tension than that which exists inside other organisations because the individual trusts are not regional offices of a parent body but independent entities which voluntarily come together to fund that central body.

Although I say it works, there are difficulties in getting everyone behind a single position (for example, on Badger culls or grouse shooting) and those difficulties are a drag on the effectiveness of wildlife conservation. There is a damaging tendency for a lowest-common-denominator approach to crucial issues. The lowest common denominator will not solve the Chronic Crisis of wildlife decline.

Also, is the spread of resources appropriate to the conservation needs of different geographies? Both Yorkshire Wildlife Trust (£11 million annual income) and Essex Wildlife Trust (£10 million) have higher incomes than the Scottish Wildlife Trust (£6.1 million) and I'll leave it to you, particularly if you live in Scotland, to decide how appropriate that is. Twenty-one of the individual Wildlife Trusts have incomes higher than Butterfly Conservation, Plantlife or Buglife, and the largest individual county Wildlife Trusts have incomes around the total of those three small but national organisations. Overall, amongst this large family of organisations the total annual income is over £200 million, a much larger sum than the income of the RSPB (£142 million) and second only to the National Trust, but it is split somewhat capriciously over 47 different pots with relatively little link to conservation need.

The 37 English Wildlife Trusts' incomes (median £4.7 million per annum) range from the Isles of Scilly Wildlife Trust (£0.3 million) to the Yorkshire Wildlife Trust (£11 million). The more one delves into the details of this, the more understandable it becomes in terms of history and the less understandable in terms of reality of current need. All these entities compete to some extent with each other for grants and public attention. Those within the Wildlife Trust family will say it works well, and I agree to some extent, but those outside the family will say that it looks rather peculiar.

The Wildlife Trusts are a much-respected and somewhat loved group of conservation organisations, but their structure embeds the fractured nature of the conservation enterprise and leads to competition between allies, duplication of resources and a weakened voice. I'm a life member of one of them because I know my support is doing good, but at heart I know it isn't doing the best good that it could.

The Wildlife Trusts exemplify within one close family what is true of the wider wildlife conservation movement – a house divided.

A house divided writ large – the tangled bank

Let us now return to *The 28*, clearly separate bodies (Chapter 3, *Who are the wildlife conservationists?*), and consider whether they are the 28 separate organisations that one would invent, from scratch, to do the job of non-governmental wildlife conservation in the UK. The non-governmental wildlife conservation sector was not designed, it has evolved. We could look on it as resembling Darwin's tangled bank of competing plants, a rich variety that is somewhat confusing, but is there a grandeur in this scheme of things? Are *The 28* quite what you would aim for if you were seeking to create a range of organisations to give wildlife what it needs?

Suppose that the chairs of *The 28* met to discuss redrawing the boundaries between their respective bodies. What would they consider? Presumably they'd agree that they wanted a new landscape that is more effective than the current one (though there would undoubtedly be voices arguing that we already have the most effective possible landscape), and that greater effectiveness would be achieved by reducing competition for influence and resources, seeking economies of scale, and allocating resources with regard not only to need but also to the chance of success. Attention to those guiding principles might then lead to discussion about how to split things up, including issues around devolution, the balance between international and UK issues, the split between taxa and habitats, and the split between campaigning, advocacy and on-the-ground action such as land ownership.

Devolution. The UK is a devolved nation, with most conservation and land-use issues decided at a national level (i.e. England, Northern Ireland, Scotland and Wales levels).

It is in Scotland that these issues are thrown into sharpest relief – not only because of the real possibility of future Scottish independence but also because of the historical decisions that have led to the current position. Scotland occupies 32% of the UK land area but has just 8% of the human population, whereas England is home to 84% of the UK population squeezed into 54% of the land. The human population density in England is six times higher than that in Scotland. If both nations were to rejoin the EU then England would have the highest population density of any EU state except for tiny crowded Malta, while Scotland would nestle in the bottom quarter between the Republic of Ireland and Croatia. Those differing human population densities, partly a result of the physical geographies of the two nations, lead to very different conservation threats and opportunities. In England population density, along with the house building and road building it demands, exerts a massive pressure on wildlife, whereas in Scotland the wide open spaces of land of low economic value seem to open up large opportunities for habitat regeneration on a major scale.

But a low human population density might mean a low potential membership base in Scotland. The Scottish Wildlife Trust, sitting in 32% of the UK land area with 8% of the UK population, amasses 5% of the total income of Wildlife Trusts in the UK. Comparing the income of the National Trust for Scotland with that of the National Trust (which covers England, Wales and Northern Ireland), NTS gets roughly a fair share of the total resource per capita (approximately 10%) but has to spread that over a third of the UK land area.

RSPB Scotland, Plantlife Scotland, Butterfly Conservation Scotland and Buglife Scotland are all nationally branded parts of UK-wide organisations. The decision on how much money to spend in Scotland, and to some extent on what in Scotland, is largely made at a UK level. Buglife is an excellent organisation, but Buglife Scotland is tiny and couldn't possibly stand on its own two feet were independence to come. The issues for a future free-standing RSPB Scotland (BirdLife Scotland?) would be different – the challenge of taking over more than half of the UK RSPB's wildlife reserves (by area) with a small supporter base.

If Scotland becomes an independent nation, and that is a distinct possibility although nothing like a certainty, then enormous change will be forced upon the NGO world. From my North Northamptonshire home, I can see many reasons for supporting Scottish independence, but since I don't have a vote on the matter I don't think about it all the time, and most of my reasons for being warm to the idea are nothing to do with wildlife conservation. When we look at the potential impacts of Scottish independence on wildlife conservation, then in the long term I think it would be a good thing and lead to stronger and more effective conservation in Scotland, but the process of splitting up the resources of UK organisations to fit the fracturing of the UK would hamper wildlife conservation in Scotland and in the rest of the UK for probably a decade.

It feels as though we are in a waiting position – waiting for a resolution of bigger questions about the future of the UK, but not dealing properly with the current reality.

International versus UK. Many UK conservation organisations spend some of their resources outside the UK. In recent years this has often meant in partnerships with European Union countries, and those will probably continue to some extent post-Brexit. However, the UK is no longer eligible to receive EU grants for joint projects, and this will seriously cramp the amount of collaboration that can be afforded. Some of the conservation work is in the far-flung remnants of the British Empire including the UK Overseas Territories (such as Ascension Island in the South Atlantic and the Pitcairn Islands in the Pacific). WWF-UK is the UK wildlife conservation organisa-tion probably most associated with international conservation, and their projects are spread widely across the globe with rather little UK activity. As I look at their website today, I am struck by the Snow Leopards, Polar Bears, African Elephants and Orang-utans that I could adopt – and then I come across a project called Wild Inglebor-ough, which seems pretty good but also pretty much out of kilter with everything else that WWF-UK is involved in. I'm not sure whether it's me who is confused, or WWF-UK.

Taxonomy. There is only one organisation among *The 28* that deals primarily with plants, and although Plantlife is brilliant in many ways, it is rather small to be looking after vascular plants and fungi, more or less on its own. I guess the Woodland Trust is a plant-based conservation organisation but it is heavily biased towards lignin. Is the wealth of marine species really well served by the organisations that are primarily marine and the efforts that larger organisations devote to marine matters? Do birds get too much attention? Will the UK's invertebrates be saved by Buglife, Butterfly Conservation and a few other even smaller bodies? I think probably not. Will the big organisations, RSPB, Wildlife Trusts and Woodland Trust, really do a good job for invertebrates? I think they might in terms of wildlife reserves, arguing for proper planning regulations and protected areas, but not so much in more fine-grained conservation work. And if everyone had believed that the bigger guys would look after the smaller invertebrates then we wouldn't have specialised invertebrate conservation organisations at all.

Approach. If *The 28* managed much of the UK land area then there'd be a lot more UK wildlife, but despite the large land ownerships of the National Trust (250,000 ha in England, Wales and Northern Ireland), the RSPB (130,000 ha), the Wildlife Trusts (100,000 ha) and the National Trust for Scotland (73,000 ha) totalling over half a million hectares, that is only about 2% of the UK land area. Even if those organisations own the best 2% of the UK for wildlife conservation, that still leaves the other 98% which has to be influenced by other means, largely through campaigning and advocacy aimed at government, although partly through advice and friendship to individual private landowners. The National Trust and the National Trust for Scotland lean heavily towards the land-management route whereas organisations such as Greenpeace and Friends of the Earth are much invested in the advocacy and campaigning route without owning land at all. Everyone else does a bit of both, or focuses on the marine environment.

As we've already seen, wildlife conservation involves difficult choices. Its success depends on enough of us deciding that it is an important aspect of human endeavour, so that it is given higher priority and gains more ground against the competing human activities that lead to wildlife loss. And once you are doing wildlife conservation you are faced with four main strategic choices about how to allocate resources, which I have labelled above as *devolution*, *international versus UK*, *taxonomy* and *approach*. Within the UK, how do you allocate your resources between different places and regions? Do you in fact focus on the UK, or should you attend more closely to the needs of the rest of the world? Should you specialise in a particular group of animals or plants, and which ones need the most help? And what sort of actions should you engage in, what sort of approach should you pursue?

Looking at the activities of our wildlife conservation organisations, it is difficult to get away from the conclusion that the marine environment, plants and invertebrates, Northern Ireland, Wales and Scotland are under-resourced in terms of their conservation needs whereas birds and mammals in England are much better off – although quite clearly not to the extent that they are thriving under current circumstances. This is understandable, since birds and mammals are well-loved parts of our fauna and most of the UK population lives in England. Yes, these are partly market forces at play (and market forces rarely deliver public goods efficiently). And the more you look, the more complicated it becomes. After all, it would be very easy to make the case that the RSPB does more for plant conservation than Plantlife does, but that is partly based on its much greater size (and greater clout and land holdings). It would be impossible to argue that if Plantlife had the RSPB's resources then plant conservation would not get a massive boost.

The organisational structure of wildlife conservation in the UK is complex, with many organisations competing for resources and influence. There is perhaps an infinite number of ways of arranging the architecture of NGO conservation bodies that would be even worse than the structure we have at the moment, and probably a similar number that would work better. The question is not whether there is a better structure of organisations, but rather, if there is a better structure, then how do we reach it quickly and efficiently? Now that is a hard nut to crack, and I think it is beyond me. And that meeting between the chairs of *The 28* to discuss the boundaries between their respective bodies will never happen. We are largely stuck with the current structure, and we must work within it to make the best of things.

Are the wildlife NGOs doing a good enough job?

Without the effective and important wildlife conservation work carried out by *The 28* things would be even worse. However, I feel that those 28 organisations aren't doing as good a job for wildlife as they once did.

Peak NGO, by which I mean the heyday of traditional, large, membership-based, wildlife conservation organisations, started in the 1980s and lasted perhaps until around 2010. During that period the NGO community achieved a massive amount through expanding their horizons, their memberships, their ambitions, their incomes and their power to do good. I think they were maximally effective back then, and maximally focused on wildlife conservation too. That was their era, that was when they looked like the cutting edge of wildlife conservation and things got done. Since then, they have become tamer, more cowed, more distracted, and have lost focus. I hope this a temporary decline, but it is a decline, and it's an unavoidable part of the problem.

There are six main problems, in my view, which mean that our current mix of wildlife charities is not operating with maximum impact.

Failure to celebrate success. *The 28* are operating in a world where wildlife is in decline, and they are trying, unsuccessfully at the moment, to reverse that decline. That might feel like a difficult position but it really isn't. The public understand that turning around a long-term decline is difficult and they admire conservation charities who pluckily take on the task. But that means we must do more to publicise any successes that are achieved. That is what will make the existing supporters feel glad that they gave their support, attract new supporters and create an atmosphere of hope about what can be done. And then it's easy to slip in the message that others, usually government or its agencies, should be doing more too.

We need more good news stories about conservation successes in order to inspire more support for even more conservation successes. These need to be brilliantly illustrated, clearly explained, but modestly told. In an overall landscape of wildlife decline, successes need to be celebrated as examples that show the way even if they cannot be counted as overall victory.

Failure to buy more land. Many of *The 28* own and manage land, as wildlife reserves. It feels to me, through the communications that I receive from such bodies, that this tried and tested mechanism for delivering more wildlife is now underused. I haven't had a request to contribute to a land purchase from major conservation bodies for ages – and yet I would willingly give to such an appeal. Indeed, I have donated to other requests from the likes of Heal Rewilding (which crowdfunds for rewilding work) and the Tarras Valley project (which is turning a former grouse moor at Langholm into a rewilded mixture of moorland, scrub and woodland).

It feels as though our wildlife conservation organisations have been seduced into thinking that they can solve all wildlife's problems by strutting through the corridors of power in Whitehall, Cardiff, Belfast or Edinburgh rather than by spending their money on soggy fields and beautiful woodlands far from the seats of power. I'm fairly firmly wedded to a mixed approach of doing some of both, depending on the opportunities that are presented at any particular time.

Failure to call out government failure. The relationship between NGOs and government and its statutory agencies is a complex one. Is government an ally or an enemy? Well, the balance changes over time and it's never all ally or all enemy, but the rule of thumb has to be that everything is the fault of government and they are moving too slowly, too inefficiently and too grudgingly to make the difference we need. The advantage with this view of life is that it is almost always true. Governments across the UK pay scant attention to the Chronic Crisis of wildlife decline, and that's why things aren't much better.

When government agencies do the right thing then they should be praised – usually with the caveat that they haven't done enough and didn't do it quickly enough – but when they don't act then they should be called out in very clear terms. We don't hear our NGOs doing this very much. They look as if they are part of the establishment rather than that they are trying to change the world. The suspicion has to be that they are afraid of losing funds from government grants and contracts and access to those highly prized meetings with government ministers and top officials. Our NGOs will be told that they are valued partners, friendly critics and of great influence – whereas it looks from the outside as though our NGOs have little influence but are given enough access to make them feel better about life.

The very language of wildlife conservation organisations about government has changed too. It always brings me up short when I hear conservationists whose salaries I am paying referring to those who work for the government departments who aren't doing their jobs as 'Defra colleagues' – what is that all about? And the same or- ganisations often regard themselves as 'delivery partners' because they'd quite like to get a contract from government to do some small bit of nature conservation action. I see a lot of NGO press releases, and they far too often express 'disappointment' over what public bodies have or haven't done, rather than a forthright condemnation of the government position.

We have a conservation movement that is too timid in its criticism of governments across the UK for their inaction and failure. Such timidity reduces the conservation movement's ability to have any leverage with political events. Vocal criticism in the media, and specifically to the millions of wildlife supporters out there, is something that politicians fear. In an argument between government saying it's doing a great job and wildlife conservation organisations saying the opposite, there is no doubt who the public will believe. Public criticism is part of the power of the NGO world – it's a strong weapon and it must be used often. It must be used fairly, but it must be used, otherwise neglecting wildlife conservation has no consequence.

Too much energy is spent commenting on what government is doing rather than telling government what it ought to be doing. Our wildlife organisations used to set the agenda for government action, they were once leaders; but now they are too often mere followers. There is an urgent need for some game-changing measures to end the wildlife crisis – these should surely come from the professional conservation organisations.

Failure to call out vested interests. It's good to talk, but the talk has to be productive. Where organisations representing different interests clash then there is a duty to explore the issues and try to resolve them. But at some stage the talking must stop; talking must not block progress. Our wildlife NGOs act as though they must persuade vested interests to become unvested – to stop wanting to do what it is clear they really, really, really want to do. This will never work – it never has and never will. You might as well try to talk Red Foxes out of eating wader eggs – it just won't happen. In advocacy you have to realise that you must jump over the heads of vested interests and go straight to government.

I remember a fairly recent discussion with other wildlife conservationists about a joint approach to a particular issue, and someone said 'But [the industry] won't like that' – as though that were the end of the discussion. It isn't the job of wildlife conservationists to trim their asks to being acceptable to the industries doing harm to wildlife. Do we suppose that in the parallel conversations within industry, anyone ever says 'But the wildlife NGOs won't like that'? Of course, they don't. Our job as conservationists is to be clever and skilled advocates to government, not to win friends in every vested interest under the sun. Yes, we should offer workable solutions to decision makers in government, but not compromised solutions that already give away huge chunks of what wildlife needs.

Pitching your 'ask' of government correctly is always a tricky decision, but treating your opponents as though they are your allies has never been a successful strategy.

Failure to mobilise the public. Our wildlife NGOs tell us, more than anyone else, that wildlife is in crisis, but they are not themselves in crisis mode. They are in comfortable mode. Thank you for your money, there's a wildlife crisis, would you like to buy some bird food now or perhaps bring your children to a sleepover? Those are the messages that come through my magazines, newsletters and emails. Why aren't the supporters of wildlife being mobilised to lobby governments for more action?

The biggest fairly recent opportunity to influence a political decision with huge importance for the environment was the Brexit debate. Now this was a very political subject but there was no doubt whatsoever that it was of huge importance to wildlife conservation, because so much of our wildlife legislation had been formed during our years of EU membership. And the writing was on the wall that Brexit would lead to watering down of that protection for species and habitats. The protection that EU membership brought with it was far from perfect, but we'll miss it when it goes, and I'm certain it will be chipped away over the next few years, at least in England. You might doubt my analysis, but let me tell you that it was shared widely across most wildlife NGOs – if you had carried out a secret ballot of conservation professionals you would have found, on environmental grounds, a massive majority for Remain. That being the case, those wildlife conservation organisations should have made a strong case for Remain – who knows, it might even have made a difference.

With the notable exception of the Wildlife Trusts (who, I thought, were the most outspoken – though that is bigging them up a bit) *The 28* remained eerily silent before the referendum in 2016. I'm sure that was through nervousness about charity law and entering into a political debate. Well, in that case, all of those organisations should cease all lobbying of politicians forthwith. I will never forgive our wildlife NGOs for being so feeble – they hid away on the biggest environmental issue of their lives. When leadership was needed, it was absent, and this was very much noticed by politicians on both Remain and Brexit sides of the debate. I remember a Remainer MP with impeccable environmental credentials metaphorically tearing her hair out over the lack of comment on the environmental impacts of the coming referendum from *The 28* to their enormous memberships and supporter bases.

Here's another example. In 2021 over 50 of the organisations making up Wildlife and Countryside Link (WCL) supported a petition which was part of a successful campaign to get the Westminster government, in its Environment Bill, to commit to a legally binding target to stop the loss of wildlife. I was actively involved in that campaign through Wild Justice and so I know quite a lot of the ins and outs of it, but I will stick to those elements that are in the public domain to illustrate my point. As part of the campaign, those 50+ organisations asked their members to sign a petition, and the petition reached 208,000 signatures in the four months of March to June. Over 200,000 people is a lot, but it's a long way short of the alleged 8 million combined membership and supporters of WCL's member organisations. As a measure of the ability of WCL's member organisations to mobilise their own memberships and supporters it looked weak. That low number suggests either that the NGOs weren't trying very hard or else that their supporters weren't really very interested in speaking out, at the cost of a few mouse-clicks, for wildlife. Either way, it's a failure.

Signing petitions is a very honourable and tested way to demonstrate public feeling about an issue, and it also gives your supporters an outlet for their feelings that is constructive and a highly respected part of the democratic tradition. I've been involved with several petitions, particularly ones set up on the UK parliament website calling for action on environmental issues. I've started some of these and promoted those of others. Petitions won't, by themselves, change the mind of governments but they are a very tangible signal of public opinion and part of the campaigner's armoury.

Our wildlife conservation organisations fail almost completely to use the official channels, the petition functions available in Westminster, Holyrood and Cardiff, to make a case directly to those parliaments. I say *almost* completely because the Wildlife Trusts, in essence, through their President at the time, did launch a Badger-cull petition to the Westminster parliament in 2016 which was debated in 2017, having gained 108,320 signatures. Instead, the wildlife NGOs tend to set up their own petitions or use other sites' software. I think this is a shame on two counts. First, petitions on the official parliament websites trigger government responses, which is always a good thing, and if sufficiently well supported they trigger a debate on the subject. In other words, they help to raise the issue more widely. Second, I fear that many petitions these days are set up with a mind more to gathering names and contact details than making a political point.

Failure to attract the right supporters. There are two types of supporter that a wildlife conservation organisation can recruit – the cause-led supporters and the transactional supporters. We're all a mixture of both, but wildlife charities get the supporters they deserve. If your communications with the public have lots of information about free parking and entry to sites, the benefits of wildlife watching for your mental health, family days out, yummy food in the café, a great magazine and how the bird food on offer is the best available to fill your garden with birds, you will tend to attract transactional members. Whereas if you talk about the problems faced by wildlife and how your work, including active campaigning, can turn things around and really make a difference, then you will attract cause-led supporters. All the wildlife charities know this, and there has been a serious switch to transactional members over the last decade, which means that the memberships of our wildlife conservation organisations now are far less committed to wildlife conservation. They may be quite keen on wildlife but they aren't that motivated to help it. This change, which in some ways was led by the National Trust, is fine if, like the National Trust, your heart is not in nature conservation, but it's a disaster if you really want to make a difference in the world. Many wildlife organisations have made a rod for their own backs by letting the marketing departments control the communications to the public, and it will take a while to get things straight again.

There has, over the last quarter of a century or more, been a move to professionalise the activities of our wildlife conservation organisations so that they are better run along more business-like lines. This has been a good thing on the whole – greater efficiency is always a good idea. But in chasing a wide variety of sources of money *The 28* have recruited less-committed members, increased the amount of merchandise that they sell, and become more dependent on statutory sector funding. None of this really aids their cause, and none of it makes them look like lean and mean conservation machines. Whisper it quietly, but our conservation organisations too often look like vested interests, communicating for their own benefit, rather than interest groups focused on wildlife recovery.

These six issues are all mutually connected. If you recruit transactional members then you spend more of your time giving them what they want and a bit less giving wildlife what it needs. You might even communicate your conservation successes less well and less often to that audience, as those members aren't so turned on by the cause or by helping to influence change in the world around them. That means that government ceases to be impressed by all those voters who support your organisation, judging that they are a rather passive bunch. And if statutory sector funding becomes more important to you then your enthusiasm for criticising government lessens and you weaken wildlife's voice even further.

The extreme poverty of the NGO world

The 28 have a collected joint income of over a billion pounds per annum, which sounds quite amazing, except that around half of that lies in the £508 million annual income of the National Trust (little of which, by my estimation, is spent on wildlife conservation), and another £84 million is in the hands of WWF-UK (which spends much of its money abroad). Of the rest, most of the income, some £400 million, resides in RSPB, the Woodland Trust and the Wildlife Trusts. The National Trust for Scotland has an annual income of £44 million, and the Wildfowl and Wetlands Trust one of around £21 million, but then all the rest are below £5 million (some much below). What this tells us is that unless the RSPB, Woodland Trust and Wildlife Trusts are doing a great job with their £400 million then most of that money will be spent abroad, on stately homes and landscapes, or on relatively small wildlife conservation projects.

The size of the pot of money to be spent on UK wildlife conservation by dedicated wildlife charities is therefore around £500 million per annum. From that sum, the wildlife NGOs have to pay their staff, provide them with toilets, pensions, computers, phones, transport and other equipment, run wildlife reserves, lobby government and recruit and retain their memberships. Essentially, a small number of wildlife NGOs are trying to ameliorate the impacts of 68 million people in the UK with a financial pot of half a billion pounds (that's £500,000,000) in an economy of over 2 trillion pounds (that's £2,000,000,000,000). Or, to look at it another way, *The 28* have about £7 per UK resident to spend on wildlife conservation. Does that sound like enough?

For comparison, there are eight charities registered in England alone which have annual incomes of more than £500 million per annum. Top of the list is the Arts Council with an annual income of £1.5 billion. A search on the term 'cancer' in English charities reveals over 1,500 individual charities, and the annual income of just the largest four of them is well over a billion pounds a year, with Cancer Research UK's annual income topping that list at £580 million.

We must face the fact that although it sometimes looks as though there are loads and loads of wildlife conservation NGOs out there, on closer examination the resources available to them are paltry compared with the resources allocated to providing other public goods. And compared with the amount spent on potentially damaging activities, the sum is tiny. Even if all were fine in the activities of *The 28*, these are meagre resources with which to help UK wildlife to thrive.

We are few

How many people in the UK really care about wildlife conservation and are prepared to act on its behalf? I do, and let me assume that you do, but how many others?

The RSPB is well known for having a million signed-up members and used to have a strapline of 'a million voices for nature'. That seven-figure total is often commented upon by politicians with a certain amount of awe.

Membership is a tangible measure of commitment because it entails spending money, but to what extent does it indicate commitment to the wildlife conservation cause? The RSPB membership has been at over a million members since 1997 but that may give a false impression of constancy and commitment.

The RSPB's membership includes nearly 200,000 children who have arrived at that place through being signed up as family members by a relative because the child has shown a passing interest in birds. We shouldn't count them as committed wildlife conservationists – although the hope is that, like me, even if their childhood interests wax and wane through teenage and early adult years, they will come back to those roots.

Aside from the natural demographic changes of births and deaths there is a lot of flux in membership – which is somewhat inelegantly referred to as 'churn' and is the flipside of retention. If you retained all of one year's members into the next year then any new recruits you enticed would lead to an increase in membership. But there is a battle for members between the wildlife charities – just look at all the recruitment requests we are constantly bombarded with in adverts, emails, magazines, and popping up uncalled-for on the internet. And the fact that the RSPB membership, just as an example, has stayed at around a million members for 25 years shows that the organisation is running to stay in the same place, just like Lewis Carroll's Red Queen. The RSPB has to recruit an army of new members, let's say 120,000, every year just to stand still in the face of churn and mortality.

My estimate of the number of committed wildlife conservationists in the RSPB's million members is around 250,000 –that's a guess, but a pretty well-informed one.

But that's just the RSPB. Wildlife and Countryside Link's 60+ constituent members have over 8 million supporters combined, so what about the other 7 million? I have news for you. There aren't another 7 million!

There's a lot of double counting in that figure because many of us are supporters of several wildlife conservation NGOs. I've been a member of 13 of the WCL constituent bodies at one time or another. I am a life member of my local Wildlife Trust as well as of the RSPB, and I have been a member of other Wildlife Trusts in addition in the past. The overlap between the RSPB and the Wildlife Trusts is massive. If there are 250,000 committed wildlife conservationists in the membership of the RSPB there will probably be a similar number in the Wildlife Trusts too – the problem is that many of them will be exactly the same people (and I'm one of them, not two of them). You won't find many members of the much smaller wildlife organisations who aren't already also members of larger bodies too – they are the same people.

Then there is the National Trust with its 5.4 million members – equal to the population of Finland, they say. The National Trust is the ultimate transactional organisation: it offers tea rooms, shops, restaurants and nice places to go for a walk, and they'll charge you to visit many of them unless you are a member. If you live near the coast, then you'd be crazy not to consider NT membership because it will save you a fortune over the year. There will often be a charming person standing near the pay-and-display machine pointing out to you that if you sign up for membership now you won't have to find the coins to put in the machine today or any time in the next 364 days and you'll be able to spend your money on delicious cheese scones as well. This is a successful financial model but it won't attract cause-led members. I cannot really believe that within the National Trust's vast membership there are many extra committed wildlife conservationists who aren't already members of the Wildlife Trusts and/or RSPB. The National Trust doesn't often act like a wildlife conservation organisation and it doesn't treat its membership as being much interested in the subject, so I think we can be sure that a strong interest in wildlife conservation is a rare perspective in that organisation.

How many committed wildlife conservation supporters are there out there? We don't know, but it certainly isn't everyone, nor is it most people, and it isn't millions and millions; my guess is about half a million. I might be wrong, but there are no figures that would tell us the answer so we have to estimate it based on what we do know. You may think that there are many more – but if there are, where are they hiding? And if government really believed that the wildlife conservation movement had 8 million supporters it might well take a lot more notice of what it said.

A comparison with another interest group may be instructive. In 2021/22 the average weekly attendance at Premier League football matches (i.e. just England, just the 20 top teams, and just football) was around 400,000, and these are people who are paying to go to the games, including in winter when it's cold, dark and often raining. A season ticket for Liverpool FC for 2022/23 costs an adult around £750–850, depending on which stand it is (reductions for old and young), and there are 20,000 season-ticket holders and a waiting list of another 70,000.

Wildlife conservation is a niche interest, and we are fooling ourselves if we think differently. The support for wildlife is mostly passive support. That is a sobering view and the consequences are great. Our favoured NGOs need to do better at mobilising the casual interest in wildlife into an active interest in wildlife conservation. The consequence for you and me is that we are a much higher proportion of the committed wildlife conservationists than we thought we were – so we'd better do our bit.

Reflection 5

This book may feel like a series of highs and lows, and if so, this chapter may be the lowest of the lows. It's bad enough to know that wildlife has declined for ages and ages, but then to be told that the state of wildlife conservation is every bit as worrying as the state of wildlife is a bit of a downer. It's been a bit of a downer to write too.

The three main ways in which we are failing wildlife, and allowing continuing chronic wildlife declines, are (1) that we don't invest enough money in wildlife conservation (this is the biggest problem), (2) we don't regulate against wildlife-harming activities effectively enough, and, less important, (3) we have a very peculiar conglomeration of wildlife charities which we fund to sort all this out.

Wildlife conservation is a difficult endeavour, not because we don't understand what needs to be done, not because it's impossible to do what needs to be done, but because it's impossible to make much headway with the resources directed at the issue. We have the puny resources of a small number of NGOs and the inadequate resources that government allows to be directed at the natural world on one side, and almost the rest of human activity, one way or another, and a largely unbothered nation, aided by some powerful vested interests, on the other side.

We wouldn't live longer than our grandparents did if our prede-cessors hadn't invented the National Health Service and resourced it appropriately. And we wouldn't understand as much about the world if we didn't all go through at least a decade of schooling paid for from our taxes. These are societal choices that were made long ago. The scale of the financial spend is important because it determines how much gets done. Slash the resources for health or education, and that aspect of our lives will suffer. We haven't made anything other than a wishy-washy societal choice to reverse wildlife declines, and we certainly haven't allocated the fairly modest resources necessary to make that happen – and so we live in a time of continuing loss of wildlife. That says a great deal about our true relationship with wildlife.

Governments, if they were at all serious about stemming wildlife losses, would use regulation much more. For one thing, it is in some senses free. Banning the sale of peat compost does not cost the national exchequer money – but it does affect those who are currently making money from selling peat-based composts, and they will kick up a fuss. So will the users of wildlife-killing pesticides and lead ammunition, and those who burn upland vegetation for grouse shooting. They will say that their incomes are affected – and they probably have a point – but it's up to government to decide who should be the winners and losers. Wildlife should more often be a winner, and we must provide the voice for wildlife.

If you were starting from scratch, no one would design the non-governmental response to wildlife decline to look like it does today. We have a strange collection of organisations who spread their resources geographically and taxonomically across the UK according to no shared or agreed rationale. They compete with each other for the same limited government funding, and for individual memberships. And they are conflicted as to whether they are there as partners of the statutory sector or whether their job is to give decision makers frequent painful pokes with sharp sticks to keep them up to their tasks. But the two best things about *The 28* are, first, that they exist to help wildlife so they are definitely and unswervingly on the right side of the debate, and, second, that they depend on us, the public, for their salaries so they are open to persuasion by us, the wildlife enthusiasts across the land.

It's surely time for a reboot. What is the way forward? That's the subject for the final chapter. Don't expect a magic bullet, don't expect it to be easy, and don't expect that somebody else will do all the work…

CHAPTER 6

What wildlife needs (and how to provide it)

What does wildlife need? And how can we provide it? That is the subject of this last chapter.

Here, I set out in very general terms what sort of future for wildlife I'd like to see. I'm quite specific on some things, but I don't go overboard on the vision thing, on the outcomes, on the ends. I'd rather not enter a Monty Pythonesque argument between the Front for the Conservation of Wildlife and the Wildlife Conservation Front about who has identified the only true wildlife-rich outcome. If you are a fan of wildlife and want to see more of it then you and I are on the same side. But we still face choices about how to engineer change, and those choices are big ones and will require a step change in how we do wildlife conservation in the UK.

Only a step change will end the chronic decline of UK wildlife. A little bit more of the same will leave us with many more years of depletion. That step change must involve greater expenditure – I can't see any way that wildlife losses can be stemmed on the cheap, although, having said that, I think delivering more wildlife across the country will be cheap in overall terms. And the more that governments use regulation rather than incentives, the cheaper will be the overall bill for a wildlife-rich UK.

I concentrate on the means to achieve more wildlife because that is what we have got wrong for many years. We don't need more vision, we need more action, and it has to be effective and game-changing action. We need to do wildlife conservation differently in future. You and I have a role to play in making that happen, but we certainly can't do it individually. We need to work together to achieve anything substantial.

Alternative visions for a better future

If a marine conservationist asked me what I would like to see in the next 5, 10 or 50 years for the seas around Britain, then I'd probably say that they are better placed to know than I am. However, if forced to have a stab at an answer then I'd say I would push for a marine environment rich in all forms of wildlife, and that will probably mean reducing damaging activities that affect the seabed and its wildlife – things like bottom-trawling and overfishing generally – and making sure that extractive industries such as oil, gas and gravel are steered away from the best wildlife sites. I might say that I guess we could best achieve this through regulating overall fishing effort and the types of fishing that are allowed in different sites, including some no-take zones, and making sure that the youthful protected-area network is made effective and expanded. I'd guess that there will be continued changes in our local marine environment due to climate change and we can't in the short term do much about them, but we'd better take them into account as the backdrop for our action. I'd hazard a guess that pollution, certainly from sewage and agricultural run-off, which is so damaging to our inland waterways, wouldn't be such a bad thing for the marine environment if only we could get it to arrive far out to sea rather than concentrated on our beaches. I'd be a bit nervous about stating it, but I'd suggest that, in my limited experience and knowledge, invasive species are not so much of a problem. That would be my rather vague agenda for action, and I reckon that if we made progress with that then in as little as 20 years we would catch more fish sustainably for human consumption, see more cetaceans and seabirds using the marine environment, and know that under the waves the seabed was healing and becoming much richer in wildlife.

I'd be interested to hear in return what an expert marine con-servationist had as their vision – would it be very different or quite similar to mine? But then the conversation ought to move on to which parts of that vision are most achievable, and what are the best ways to go about achieving them. This would have to get into the difficult area of prioritising action, because resources will be short. Some sort of SWOT analysis (strengths, weaknesses, opportunities and threats) might help, and might lead us to press on with some parts of our vision at the expense of others simply on the pragmatic grounds that we feel we can make the most difference that way.

If my imaginary expert marine conservationist were to ask me what my good and achievable outcomes for wildlife on land would be then I'd feel on slightly firmer ground, but there are still infinite numbers of choices to be made. One of the intellectual debates of the current age is between land sparing and land sharing, and this is very relevant to the UK situation where 70% of our land area is farmed, and wildlife richness has fallen dramatically on that farmland for decades. The land sparing approach is to say that taking some land out of agricultural production and dedicating it to wildlife conservation will achieve more than sharing, which attempts to find ways in which wildlife can survive and thrive alongside agricultural production. Sharers would say that there is no mechanism to convert enough privately owned farmland into wildlife havens to make much difference but that if only we could get a grip of the way that billions of pounds of public money are spent on farming each year, and throw in a bit more regulation with it, then we can advance wildlife conservation in that 70% of our country.

Faced with a choice between sparing and sharing, which should you choose? Well, I think you should reject the premise of the question – it's not a choice that is on offer, and in any case there probably isn't a one-size-fits-all answer. We are in the current position of massive wildlife losses in the UK because we haven't spared well enough and nor have we shared well enough – neither has been a great success and we need to up our game in both areas. And I think that is the key point – the status quo in wildlife on the ground is not acceptable to me, nor to others who care about wildlife, and that biological outcome is the result of past and recent approaches to wildlife conservation – so a status quo in approach is not acceptable either. All it can promise is continuing declines.

I'd like to see the wildlife in the countryside return to the richness of my youth of 55 years ago, but I'd also like our protected areas and wildlife reserves to be bigger and richer in wildlife – I'd like both, and we need both if we are to address the long-term crisis in wildlife decline. So I'm not going to opt for sharing over sparing, or vice versa, because I'll grab whatever is practically achievable, and support any individuals or organisations who have a good plan to move forward in either way.

This false dichotomy is essentially a re-run of a debate which raged, with some passion, in the little world of the RSPB when I became conservation director in 1998. Should we concentrate on wildlife reserves as a tried and tested, and long-term, mechanism for saving wildlife, or should we concentrate on winning policy change that would affect the whole landscape but could be reversed by future policy changes? The way that debate was resolved getting on for 25 years ago was to decide that we'd continue to do both. And I'm pretty sure that was the right decision back then, as there were strong signs of progress on both fronts. At the end of my almost 13 years in the job I looked back at those years and saw three major pieces of environmental legislation under the Labour government (Countryside and Rights of Way Act 2000, Climate Change Act 2008, Marine and Coastal Access Act 2009) and signs of progress on farming grants as testimony to the advocacy route, alongside a wide range of expanded and new RSPB wildlife reserves as testimony to the land ownership approach. But at that time, one year into a coalition government in Westminster that was taking a hatchet to statutory wildlife conservation funding and structures, any route depending on persuading government to act differently looked tougher than ever (not that it is ever easy). Then I would have shifted the balance of RSPB resources towards more advocacy strength in Wales and Scotland, where there seemed to be more scope for progress, and at the same time towards greater investment in RSPB-owned wildlife reserves generally, where there is always scope for progress.

And what now? Since 2011 we have simply not seen much progress on the wildlife conservation policy agenda. This has been true not just in England, where the coalition and Conservative governments have been antagonistic to wildlife conservation and unreliable at meeting any of their promises on the subject, but also in Wales and Scotland. Wildlife conservation is severely weakened by Brexit, which is precisely what many Brexiteers in the land-owning classes wanted, and so my mental pendulum on the sparing/sharing continuum has swung further towards sparing, and ownership of land by conservation organisations, as a pragmatic rather than an ideological choice.

We need to do better on many fronts.

Getting power over land

The key to a wildlife renaissance in the UK is influencing land management. It's as simple – and difficult – as that, but it's good to be clear about what is needed.

At a very small scale, the management of my house and garden determines to a very large extent the fate of the wildlife there. Because I feed birds in the winter there are more birds in my garden than there otherwise would be. The Ivy is abundant and has spread across the shed roof because I let it. The Black Garden Ants survive because I tolerate them, and the House Mice are gone because I didn't tolerate them. In years to come when someone else lives here things will undoubtedly change. The new owners can paint the (currently green) front door purple if they like and can dig up the grass and get rid of the brambles, but for now I'm in charge. Not in complete charge because I can't reintroduce Lynx to my garden very easily or very effectively and I can't stop the local cats passing through, nor can I stop the climate changing around me or prevent nitrogen raining on my plot. But I can let the Daisies grow in the lawn and I will see the difference that makes. Doing my bit, on my own small plot, has a high feel-good factor – I can see immediate results.

However, I know that I'm making, quite honestly, a tiny difference. Whatever I do in my garden may be high impact just inside that small area but I fully understand it is low impact on wildlife conservation in the round. This would be alright if I could persuade myself that if everyone did what I do then we'd be saving the UK's wildlife, but even if every garden in the country were to get this sympathetic treatment, it would make little difference to the fate of most UK wildlife – we'd still have an almost undiminished crisis of declining wildlife. The decline in farmland plants, insects and birds would be only very marginally affected, the marine environment would not be improved, very few declining species would get a boost, damaging developments would still go ahead, and the horsemen of the ecological apocalypse would still be doing their worst over most of the countryside. Doing my bit is worth doing, and I'll do it (as I expect you will too), but we need to make more of an impact.

It's a shame that the relatively easy things to do – like eating less meat and dairy, flying a lot less, driving quite a lot less, making informed decisions as a purchaser, planting a tree, being an organic gardener – are all worth doing (and I do them) but they aren't the solution to the problem that is a chronic, ongoing steep decline in UK wildlife. They are really ground-zero, a starting point. Please do them – use less plastic, buy renewable energy, wear your clothes until they wear out (and then get them mended), recycle, turn your thermostat down, don't lash out at wasps – but we need to do much more to make a real difference to UK wildlife. We need to change the system and roll out wildlife conservation measures on a much larger scale. That's what I will concentrate on in the rest of this book.

I have seven proposals for how we can make a big difference for wildlife. They are modest but game-changing proposals if we can secure action on them. The first four are for step changes in government action, and the last three are for what you can do to make them and other really good things happen. Achieving my first four proposals (or anything else on a similar scale, because perhaps you have better ideas than mine) will not be easy, or quick, or a bundle of laughs. I'm afraid that changing the world is hard work, and has quite a lot of frustration attached to it. It's always been like that and I guess it always will be. But you have a part to play.

Proposal 1 – Let's have an effective protected-area network

An effective system of protected wildlife areas is at the heart of wildlife conservation.

The importance of protected wildlife areas on a global scale has been recognised by the IUCN and others in promoting a target of 30% protection of the Earth's land and sea surface for wildlife by 2030, with the expectation that many of these areas will currently be remote and relatively pristine in character. In the UK, former Prime Minister Boris Johnson jumped onto this bandwagon and promised that 30% of the UK will be protected by that time too. Demonstrating the UK government's weak grasp on the subject, Johnson claimed that 26% of England's land surface is already protected, so all we need to do is add a measly 4% and the job is done.

It would be great if the wildlife value of 26% of the UK (or English) land area were already properly protected, but this is far from being the case. The real figure is less than half that area, around 11%, because that is the area which is currently designated with some form of statutory (although not necessarily effective) wildlife protection. To get to 26% you have to cheat by including all the land designated for landscape reasons but which confers no specific protection to wildlife – National Parks, Areas of Outstanding Natural Beauty, National Scenic Areas.

The 11% is made up of Local Nature Reserves (LNRs, which are areas recognised by local authorities, mostly small sites), National Nature Reserves (NNRs) and SSSIs/ASSIs (which are national desig-nations, originally UK, now devolved, the bulk of the protected areas), and SPAs and SACs (mostly SSSIs but with greater legal protection). SPAs and SACs are designations derived from our 45-year relation-ship with the EU; they were once called Natura 2000 sites but are now called the National Sites Network in Brexit-voting England and Wales and European Sites in Remain-voting Scotland (I have no idea what Northern Ireland will call them).

The 11% of our total land area with statutory wildlife protection represents a treasure chest containing most of the very best places for wildlife in the UK. We need to concentrate on that 11% rather than chasing illusory offers of 30% land area. Better, I believe, to strengthen the protection of the sites already identified as deserving protection because of their wildlife riches than to aim for a larger area with inadequate protection.

The LNRs have weak legal protection, NNRs and SSSIs have quite good protection but their management could often be very much better (NNRs are mostly SSSIs but ones owned and/or managed by statutory agencies), and the sites formerly known as Natura 2000 sites have pretty good protection but that protection is under threat, particularly in England, in the post-Brexit era. So, what should we do?

We should upgrade many LNRs to SSSIs and transfer the levels of protection given to the sites formerly known as Natura 2000 sites to all biological SSSIs. The area covered by those protected sites should increase from the current 11% of land to a modest 15% by 2030, and all should have management regimes which protect and enhance their wildlife. This modest proposal would allow the Brexiteers to make good their promises that no environmental protection would be lost as a result of us leaving the EU and it would, overall, upgrade protection for some LNRs and many SSSIs (those which do not already overlap with sites formerly known as Natura 2000 sites). Yes, it represents more protection for wildlife – there is a wildlife decline crisis going on after all.

As part of strengthening and expanding the protected area system we should confer SSSI status on all ancient woodlands which are not currently so protected.

The whole business of proper and adequate management for SSSIs is an area that has fallen into disrepair under the current Westminster government despite great leaps forward under the CROW Act of 2000. A wildlife-rich grassland won't stay a wildlife-rich grassland unless it is grazed appropriately and is spared from herbicide and fertiliser. Neglect or mismanagement will lead inexorably to wildlife declines on a so-called protected site. Progress was made between 2000 and 2010 after the passing of the CROW Act and before a change of Westminster government, but that has now stalled badly and even gone into reverse under this administration. We need to recapture the progress made in the first decade and lost in the second decade of this century. An English SSSI network of which only 38% is delivering adequately for the wildlife it is designed to protect needs urgent attention. If that means spending money, then we should spend money.

In theory, Natural England keeps a close eye on English SSSIs and reports on the state of those national wildlife assets, but due to cuts in resources they don't. I had a look online at the SSSIs close to me and found a shocking neglect. My local patch of Stanwick Lakes is one of nine SSSI units in the Upper Nene Valley Gravel Pits SSSI; one of these (the smallest by area) is in Favourable condition, the other eight are in Unfavourable condition, though half of those (including my patch) are said to be Recovering. I wasn't pleased to learn that eight out of nine units of my favourite local SSSI were in Unfavourable condition, but at least this was based on a recent visit (August 2020). My nearest patch of SSSI chalk grassland, at Twywell Gullet, very close to the A14, is deemed to be in Unfavourable Declining condition because of lack of grazing and scrub management, and that is based on an assessment from November 2017. The quiet wood, Glapthorn Cow Pasture, is in Unfavourable Recovering condition. Or is it? This assessment is based on a visit in July 2011 and the Recovering assessment is based on the Wildlife Trust management plan for the site but not, it seems, on an assessment of the site itself. During that period the Nightingales have disappeared. Titchwell Meadow is regarded as Unfavourable, No Change because of inappropriate weed management (I wonder what that means), but that assessment is based on a last visit made in July 2009.

Just locally to me, the picture emerges of Natural England assessing sites as Unfavourable but not keeping tabs on them or enforcing improvements, and that is repeated all over England. It's a broken system under the current Westminster government which was working perfectly well when this administration took over from Labour in 2010. Clearly it's not good enough, clearly we are getting poor value for our taxes, and clearly we need to press government to invest in the simple things that will lead to wildlife recovery.

Protecting and properly managing wildlife-rich sites is fundamental to maintaining wildlife in this country – it's no more a luxury than maintaining the roads network or sewerage systems (both of which are suffering too, I notice). No politician has ever said 'We're going to let protected wildlife areas go to pot', but that is what is happening. I can't quite understand why the wildlife conservation movement is so supine in letting this pass without much comment or opposition.

Proposal 2 – Increase public land ownership

No, they're not making it anymore, and the people who own land have very, very rarely given it away. Land ownership in the UK can be regarded as sclerotic, with much land having remained in the same hands for hundreds of years, particularly in the uplands.

In my adopted ceremonial county of Northamptonshire, the three largest private landowners are the Marquess of Northampton (of Castle Ashby, 6,000 ha), the Earl Spencer (of Althorp Park, 5,300 ha) and the Duke of Buccleuch (of Boughton House, 4,500 ha). Their aggregated land-holding amounts to about 7% of the county and compares with about 3,800 ha of land notified as SSSIs, 4,000 ha managed by the Forestry Commission and 2,900 ha held by the Diocese of Peterborough. The forbears of those three gentlemen were respectively sixth, second and largest landowners in Northamptonshire in 1872 when the first complete land register was compiled.

The Duke of Buccleuch is also the largest private landowner in the UK as a whole, with 110,000 ha (in 2001) including large estates in Scotland. If one adds up the land-holdings of all 30 UK dukes they come to over 400,000 ha – about the size of Somerset; but the now-devolved state forestry service owns or controls much more – around 890,000 ha.

There was no Forestry Commission until the First World War made us value home-grown timber, partly because home-hewn coal needed wooden pit props. The Forestry Commission was brought into existence by Lloyd George's Liberal government in 1919. State forestry went from no land ownership to being the largest public landowner in far less than a century, and was set up as a pragmatic response to a crisis. Timber is a private good which can be sold for money, and yet 100 years ago it was thought that the most practical way to ensure that the public need was satisfied was not to prevail upon all those dukes, marquesses, earls and less noble landowners to grow trees but for the state to acquire the land and do it itself. We should now do something similar for wildlife, as an equally pragmatic response to a current crisis.

Wildlife is a public good with no traditional market value, and yet we aim to stem a long-term and continuing loss of wildlife through exhortation of private landowners to do more, and through systems of incentives which barely seem to scratch the surface of the need. Why don't we just buy up land ourselves and manage it for wildlife – as we did for forestry a century ago? Until recently, and for several decades, £3 billion per annum was spent on a system of support payments and grants to private enterprises, mostly farmers, and yet wildlife has continued to decline catastrophically on farmland. A small fraction of that sum spent on land acquisition and management would have a major impact on wildlife in the UK.

Wildlife recovery needs an urgent but temporary cash injection to secure public ownership of land in order to create a network of Knepps and Hope Farms and to expand existing areas of heathland, meadows and wetlands across the country. Management of such sites would probably be taken on by willing NGOs, who could compete with each other for the right to manage such areas. Let's say £500 million per annum for a decade – that would make a real and lasting difference. It would change the fate of wildlife in the UK.

Half a billion pounds a year is roughly equivalent to the spend of *The 28* major wildlife NGOs in the UK, but this sum would all be channelled into buying and managing land – the most effective action, which is currently neglected by government (except for forestry). The net cost to the public would be much less than £500 million each year because there would be savings on payments to the private landowners whose land was being bought – state ownership removes, for ever, the need to pay subsidies and grants to nudge private individuals into action; instead, the state becomes the willing landowner and does it itself.

Such a modest proposal is unlikely to find great favour with farmers and other landowners, but that's because they are a vested interest, with an eye to their own profits rather than the public good. For decades presidents of the National Farmers' Union (NFU) and the Country Land and Business Association (CLA) have presided over tumbling wildlife populations, saying that they don't want to be park keepers and undermining the effectiveness of government wildlife schemes. So it will come as a shock if the public insists that governments become land-owning wildlife deliverers, themselves. This wouldn't be theft of land, for one would look for willing sellers, but I wouldn't rule out keeping an exceptional power of compulsory purchase up the public sleeve, just as in other areas of government policy.

Public ownership of land can be justified on multiple grounds. For instance, the land could deliver some timber, some lead-free game meat, cleaner water supplies, lower flood risk and carbon sequestration all in the same places – as well as much more wildlife.

Let's go back to a consideration of all that forestry land we already own. In addition to acquiring more land for wildlife we should dedicate 10% of the existing forestry land, quite possibly the 10% least suitable for timber production, to wildlife management. Personally, I would have a lot of confidence in foresters doing a good job on this front if they were given the clear instruction to do it.

My general point is that, when looked at with fresh eyes, it is bizarre that government opts out of land ownership as a means of delivering wildlife recovery. If we had governments across the UK that were really thinking about how to regenerate our wildlife, they would simply look at the success of wildlife reserves managed by private individuals and wildlife conservation organisations, and the success of projects such as Knepp and Hope Farm, and compare them with the haemorrhaging of wildlife from the wider countryside, and quickly recognise that we need more of the former for wildlife recovery. The hard work has been done by others. Government simply needs to invest in success. The amount of money involved is small. It really is a no-brainer.

Proposal 3 – Rewild the uplands

Rewilding is the trendy new part of wildlife conservation, and that means that there is much talk of it and quite a lot of action actually doing it. This is very good news provided rewilding doesn't replace all of the necessary and effective other parts of wildlife conservation. And let's just whisper that some rewilding was happening long before the phrase was coined, as many species reintroduction projects, and many habitat restoration projects of the past, would now be classed as rewilding. However, the current wave of enthusiasm for rewilding takes us much further towards restoring natural processes, otherwise known as ecology, on a larger scale and with the hope and expectation that the result will more often be a closer approximation to a pristine undamaged landscape (even if in only relatively small patches).

We need more rewilding, and the place to do it on a large scale is in upland areas which are of low economic value and whose wildlife value is way below its potential through centuries of unsympathetic management. Much is already happening, led by individual landowners as well as by wildlife conservation organisations. Scotland has seen much more upland rewilding than England and Wales, and leads the way with the restoration of natural watercourses, a more natural tree line, and in some places species reintroductions too.

Surely the Welsh and English deserve a bit more rewilded land on their doorsteps, and surely our upland National Parks are the places to do it. There must be room in the Exmoor, Dartmoor, Brecon Beacons, Snowdonia, Peak District, Lake District, Yorkshire Dales, North York Moors and Northumberland National Parks for some serious rewilding to be done. How do we make it happen?

Rewilding on a significant scale will take resources and time, but we ought to start soon. I would give each of the nine National Park Authorities mentioned above a government grant of £100,000 per annum for three years to come up with their individual bids for further funding to carry out rewilding in their National Parks – that would be a modest £2.7 million investment. The nine bids would be assessed by an expert panel over a six-month period and the findings of that panel on value for money, practicality and impact would be published, and then the bids would be put to a vote of the UK population. The future funding would be substantial – three successful National Park Authorities would each receive £50 million per annum for 20 years to establish a programme of permanent rewilding in their National Parks.

The investment of public money would amount to £3 billion spread over 20 years – that's what we have given the farming industry every year for decades whilst wildlife has disappeared from the countryside. And the impact on three of our National Parks would be immense – seen not just in increased wildlife but also in cleaner water, reduced flood risk, higher carbon storage and enhanced landscape value. Knackered upland landscapes, overburned for grouse shooting and overgrazed by sheep, would be transformed. It would be hard luck on the unsuccessful National Parks, but maybe success on the ground in the chosen areas would free up more money, and in any case there has to be some competition for scarce resources.

Such a challenge pot would promote lively public discussion about what National Parks are for, and the changes on the ground would be noticeable and dramatic. This would revitalise the role of National Park Authorities – suddenly they would have a leadership role, willing partners would beat a path to their doors, and the successful ones would have very significant money to spend. That's a recipe for change.

Proposal 4 – Get tough with unsustainable farming

We can't buy it all, and even under my rather modest three proposals above the vast majority of current farmland would be unprotected by designation and still owned by private individuals. I'm not for turning the whole of the UK into a nature reserve, but neither should we be satisfied with the current scale of wildlife losses on that remaining land.

Our highly productive industrial agriculture comes with massive costs – financial costs, societal costs and environmental costs, of which losses of wildlife are one small part. If we are to reduce run-off of pesticides and fertilisers into watercourse, cut down greenhouse gas emissions from livestock and loss of carbon in soils, improve the welfare of livestock and poultry, and put some of the wildlife back into the farmed countryside, then we need big changes in practice. How do we, as a nation, get a better deal from the farming industry?

We are sometimes told that we can, as individuals, influence farming practice in the UK. There is some truth in that but it is a very weak mechanism indeed. I buy a weekly organic box (recyclable cardboard box actually) of vegetables and feel that I am making a small difference, but I know it is a tiny difference. About 3.4% of UK farmland is farmed organically and that area hasn't expanded over the past couple of decades, and organic farming is not perfect for wildlife in any case. It's taking a very long time to have an organic revolution in UK farming.

Maybe the much-talked-of switch to a diet with far less meat and dairy in it will change land use in the UK, and to the benefit of wildlife? I am part of that switch in that there is little meat in our freezers these days, and most of it is wild venison. But my dietary shift is mostly on the basis of climate change and animal welfare considerations, and I don't think it will have much impact on land use, nor will it have much impact on wildlife.

Although approaching half of the wheat grown in the UK is fed to animals, even if the UK turned vegan overnight we wouldn't see much change in land use. Wheat is an international commodity – it can be exported and already is. If the domestic demand for wheat-fed livestock disappears from the UK our farmers will sell that wheat to the world market. The UK is a good place to grow wheat: the Romans were interested in our wheat, and the last 2,000 years have simply reinforced the certainty that wheat grows well here. And if we don't grow wheat then we'll grow another intensively managed, wildlife-unfriendly crop instead – not plant a woodland.

In a similar way, the UK exports about a third of the sheep meat we produce and imports another sixth of our domestic production. I don't know, and I suspect that you don't either, whether if you eat fewer lamb chops you will be reducing imports, increasing exports or reducing UK production, and only the last of those three options can conceivably make a difference to UK land use.

We can pay for good practice through environmental grants such as those that create Skylark patches at Hope Farm or fallow plots where Stone-curlews can nest, but in general such approaches have failed to deliver good value for the taxpayer because they have been voluntary and take-up has been low. This is still a mechanism in which I have some faith, but that faith has been much diminished over the years as I have seen government continually give in to farming interests, which, understandably but not admirably, want as much money as possible for doing as little as possible.

After all these years of environmental loss through the operation of the farming industry in a largely free market, we need to employ more sticks and withhold some carrots. As a taxpayer I greatly resent paying for the wildlife-depleted countryside that now surrounds me. Why am I forced to reward such massive failure? It is time for changes to be imposed on agriculture rather than negotiated. We need governments to recognise the role of effective regulation and stop thinking that the state can reduce the ills of farming through using my taxes to pay farmers under ineffective voluntary schemes. Too often it is land-owning ministers chatting to other landowners who come up with what the rest of us have to pay for through our taxes. The voices of taxpayers, let alone the voices of wildlife, or the voices of the climate or animal welfare, are very muted in those discussions.

There is a glimmer of hope that something moving towards that model might emerge in England as part of the post-Brexit revision of farming policy – but it is only a glimmer. Defra has moved slowly, haltingly and inefficiently towards a new model of public support which recognises that things must change, but regulation is a very small part of the future package. The current government is far too close to landowning and farming interests, and that means, alongside a blind spot on how to engineer change in an environmental crisis, that they eschew the role of regulation when it comes to farming. Until we have a new deal where farmers must adopt better practices, and will suffer the consequences if they do not, then wildlife loss and environmental degradation will continue.

The failure is a failure of political will over the years, from both major political parties in Westminster as well as in the devolved nations. We have a dispersed industry, farming, which has more public appeal than many other vested interests and much better public relations, which doesn't want to give an inch, and a political establishment that never makes farming deliver for the public good. That is a failure of politics.

I guess this one may have to wait until we see how badly the Defra plans fail, and perhaps until there is a different administration in power. That's a shame, but you can't fix everything at once, however good that would be.

Proposal 5 – Get political to change the system

The four modest proposals outlined above are my favoured routes to a wildlife renaissance in our countryside. I'm not totally wedded to the details of these specific proposals, and you may have better alternative ideas, but I am convinced that changes on this scale are needed and that they cannot be achieved without significant government investment, increased public land ownership and a greater willingness of government to regulate what private individuals do. I suspect that your proposals would have to share those characteristics for me to think they had a chance of success. We have to focus on game-changing proposals, not fire-fighting ones.

What is the mechanism to deliver those changes? I don't think we can wait for politicians to come along with similar ideas all by themselves – the need is too urgent for a wait-and-see strategy. That means that we have to advocate for change. Wildlife conservation is political. We've seen that to stand much chance of delivering significant change requires prioritising wildlife conservation over other things – lengthening the third leg of the sustainability stool. You can't get much more political than that.

I might well write to my MP about my ideas, I might even give him a copy of this book, but I will freely admit that my letters aren't going to deliver a better world for wildlife on their own, for I am but one voter out of over 80,000 in his constituency and he won't hear from many others along similar lines. I am also just one of about 48 million registered voters and he is but one of 650 Westminster Members of Parliament.

In fact, I already often write to my MP. I do this because I am absolutely sure that various businesses and interest groups are nagging him all the time about what he should support and what his government colleagues should do. If wildlife lovers are silent then the overwhelming clamour in politicians' ears, whether MPs, Members of the Senedd in Wales, Members of the Scottish Parliament or

Members of the Legislative Assembly in Northern Ireland, will be from those who do not give wildlife much priority at all.

I need to stand alongside others so that our disparate voices can add up to a clear message to politicians in and out of government about how they can please us and stand a better chance of winning our votes. Where are those others? They are my fellow members of wildlife charities, and one of the roles of those wildlife charities is to bring us together to be a more powerful voice for wildlife, as well as using our money to do practical good on the ground.

I'm not expecting the chief executives and chairs of *The 28* to read this book and adopt my thinking wholesale, but what I do expect of them, in return for your and my support, is to give us proposals for a wildlife renaissance that we can help them promote to government. *The 28* must be our route to influence government spending and action. If Plantlife comes up with the best set of achievable proposals then I'll willingly write to my MP about them, but I'm really looking to the RSPB, the Wildlife Trusts and the Woodland Trust to make the running, and quite honestly, I'd be surprised if the RSPB didn't do the best job – but I'd like to see. Wouldn't it be great if *The 28* worked much more closely together on advocacy, and instead of each asking me to contact my elected politician on different matters, they came together with some agreed campaigns.

I want to support wildlife conservation organisations that not only make a very practical difference on the ground through their own actions but also make a very large practical difference through campaigning to get more government action for wildlife.

That's what I want, because I think that is what wildlife needs. How do we get it?

Proposal 6 – Choose the best NGOs

Our wildlife conservation organisations are our best bet to get more wildlife conserved because *The 28* can, between them, buy land, reintroduce species, advise willing landowners, investigate wildlife crime, influence the media debate, mobilise hundreds of thousands, perhaps millions, of people and speak truth to power. Put like that, it sounds obvious, I hope, but I believe it is such a blindingly obvious fact that most people can't see how uplifting and positive this is. How do you give wildlife what it needs? You support wildlife conservation organisations.

Your moral support is worth something, but not very much. Wildlife isn't going to get what it needs by us all sitting at home and thinking 'Aren't the Woodland Trust doing a good job. Well done them!' No, if we think that, then our support for the Woodland Trust should be active and financial – yes, particularly financial. Volunteering is a form of financial support, but there is nothing like a cheque in the post, a direct debit or a legacy to give wildlife conservation a boost. Wildlife conservation is starved of the funds that are needed for wildlife to prosper, and there's no getting away from the fact that – at least to start with – it will be wildlife enthusiasts like you and me who will fund much of the work.

We have a poor government response to the chronic decline in UK wildlife but an eager, skilled (though far from perfect) bunch of wildlife charities who are taking on much of the responsibility for delivering effective conservation action. They face a difficult task, but luckily for us we are paying their salaries and so we have some say in what they do. We should use our power more. We get the wildlife conservation organisations we deserve, and we've got some quite good ones, but let us put our energies into making them much better.

This may seem like a modest aim – after all, they only spend about £500 million on the task each year. Well, modest though it may be, increasing that half a billion to a billion, and getting it spent a little bit better – let's say 20% better – will make a big difference on its own. But it is their ability to speak to governments that is a potential major strength, and that is where we really need them to be successful. You and I have tiny voices and are rarely asked to meetings with ministers but *The 28* are in those meetings – and they are not just the voice of wildlife but the voice of wildlife supporters too.

Think of yourself as an investor in wildlife conservation. Where is your money best invested? Forget about being loyal to the RSPB, the Wildlife Trusts or the Woodland Trust – realise that your loyalties are to wildlife. Think not of the face of your local wildlife reserve warden but of the plight of wildlife around you. Be prepared to sell your met-aphorical shares in one organisation and reinvest in another if that's where you think your cash will produce the biggest wildlife dividend.

Many of us are members of a whole slew of wildlife conserva-tion organisations, partly through inertia. The best way of rebooting your relationship with them, and getting into the mindset of a wildlife investor, is to cancel all your direct debits and standing orders and start with a clean sheet of paper. How much money have you just saved? I suggest you double that sum, please, and then get into the investor frame of mind.

I am holding a crisp £20 note from the Bank of England and it's the first time I've looked closely at one. J.M.W. Turner looks quite a dandy now that I study him – but it is a self-portrait so maybe that is how he would have liked to look rather than how he actually did appear. Its serial number is CA12366706 – maybe you'll have it in your wallet one day. But what if I were deciding how to allocate it to the best possible wildlife conservation organisation in the UK – where would it go?

Some people have thought about optimal investment in wildlife conservation in the past, but I think that you, the wildlife investor, can do even better. The Environmental Funders Network publishes an analysis, every few years, of opinions (from the environment sector, so a little incestuous, though supposedly well-informed) of the best environmental organisations. The top three organisations, scoring at least 20 votes, in their latest (2021) assessment were the RSPB (22), Client Earth (21) and the Wildlife Trusts (20). Such a ranking has lots of drawbacks, and doesn't rate much higher to my mind than the accuracy of horoscopes, but it is interesting. The RSPB with its £142 million per annum income unsurprisingly comes top of the list but the likes of Plantlife (£4 million, 7 votes), Buglife (£1.6 million, 6 votes), Rewilding Britain (£0.8 million, 5 votes) and Wild Justice (£0.2 million, 4 votes) catch the eye too. As do the National Trust (£508 million, 9 votes) and WWF-UK (£84 million, 5 votes) for a different reason.

If influencing land management is the most important thing in delivering more wildlife, and I believe that it is, then we need to invest in organisations that can do at least one of two things: acquire and manage land themselves or ensure that government policies and practices produce wildlife-friendly land management. My personal choice, already exercised, is to support at least one of the biggest three wildlife conservation organisations most engaged with UK wildlife (the RSPB, Wildlife Trusts or Woodland Trust, and definitely not the National Trust or WWF-UK) and maybe a couple of the much smaller ones too.

Of the 'Big 3' I would, it may not surprise you, pick the RSPB every time because of its strong science base, because of its land management skills, because of its UK and national (England, Wales, Scotland, Northern Ireland) rather than regional/county/local focus, and because I look at its conservation successes and they seem to me to outrank those of either of the other two large organisations. The Wildlife Trusts are a house too divided, with not enough national advocacy clout, and although I think that my local trust is above average I have even greater admiration for the Scottish and Yorkshire Wildlife Trusts – but joining a local organisation not local to me seems an odd thing to do. I have warmed to the Woodland Trust quite a lot while writing this book, and maybe you will too when you view it as an investor, but for me it is still miles behind the RSPB in value for money and effectiveness. But the choice is yours – it's your investment. And however much or little you can afford then you should think very carefully about where to invest.

If you invest in a small organisation too then I have a lot of time for Buglife, Butterfly Conservation, Open Seas, Plantlife, Rewilding Britain, Trees for Life and Wild Justice (of which I am one of the founders).

So, perverse as it may seem, the best way, the very best way, to support wildlife conservation is to narrow your support to your most favoured wildlife conservation organisations by resigning from quite a few of them. But remember, this is predicated on you increasing your overall investment – which might be in terms of membership subscription and/or one-off donations. Please don't count your purchases of merchandise as real support, as the £20 you spend that way is probably only worth a couple of quid once all the costs involved are taken into account. The aim is for us all to give greater support to wildlife conservation, and to direct that support more carefully to where it can do most good.

You will find, because you are unused to being an ice-cool non-emotional investor in wildlife, that the most difficult part of this process is stopping supporting some organisations. In the words of Neil Sedaka's signature song, 'Breaking up is hard to do'. But you should steel yourself to the task and ask yourself not what the impact will be on a wildlife conservation organisation but what the impact will be on wildlife conservation. Where would our declining species want you to invest? What does wildlife need you to do?

When you end your support for an organisation, please write to the chairs of trustees (not the chief executives) of the organisations that you have left and tell them what you've done, why you did it, and which organisations you chose to support instead of theirs. I promise you, if you did this, it would be one of the most effective, cost-effective and low-effort messages that you could possibly send. And send it, as a nicely written letter, to the chair because they don't get many letters, will have time to think about it, and are very likely to talk to other trustees and the staff about the messages coming through.

I took another look at J.M.W. Turner and I almost thought I saw him nod his approval.

Proposal 7 – Be an active investor

Investing in the best wildlife conservation organisation is the best way for you to make a difference. These are the professionals in their field and know what they are doing. We have some good organisations which have a long history of achievement of which they can be somewhat proud. But all could do better, and need to do better if the chronic decline in UK wildlife is to be reversed.

Once we have acted like investors in wildlife we should monitor how well or badly our investment is performing. It's not as though we never hear from wildlife conservation organisations: they send us magazines, newsletters, social-media messages, annual reports – and in many cases we take a closer look at our investment when we visit a wildlife reserve or attend a local group meeting or even perhaps an AGM.

All organisations monitor the number of their supporters and how much money is rolling in but they don't have very sophisticated means of assessing whether their members and supporters are happy with them and what they think of their work. Rarely are you asked what you think – it's almost as though they are more interested in your money than your views.

As an investor, you should expect to see a return. That return won't be in the form of an immediate resurgence of wildlife in our seas, farmed countryside and hills, but every month, every year ought to show signs of progress being made. If it isn't clear to you that this is happening then write to the organisation in which you are investing and ask them how they think things are doing, pointing out that there are other wildlife organisations in which you can invest your money. Again, I'd write to the chair if I were you, because that way it is most likely that both the chief executive and other trustees will see your original letter (or email) and the response and be able to judge what is happening.

Ask your chosen organisations how they plan to be game changers rather than firefighters. Ask them how they plan to engage supporters like yourself in political campaigning and what their most promising or successful projects are right now. Ask how you can hear more about their conservation work and less about things that they are trying to sell you. Ask how much land they have bought in the last 12 months if that is a part of their portfolio of action. Ask what their view is on rewilding and how they feel about Pine Marten and Beaver reintroductions. Ask specific questions and expect specific answers – if you don't get an answer, or you get a poor one, then write back quickly and press the point while reminding them that you have invested in their conservation work and are seeking information on whether or not you should continue that investment.

Don't allow yourself to be fobbed off by being directed to your investment's annual report and accounts. These are generally dreadful documents for the conservation investor because they are full of financial detail that makes almost no sense at all to the wildlife enthusiast but have just a small selection of anecdotes about conservation successes. Land-holding conservation organisations should all be telling us how wildlife is doing on their own land-holdings, but you will find that many are unable to do so, and that will speak volumes as to whether they have their eye on the task or not. I'd love to know the fate of farmland wildlife on the farmland owned and managed by *The 28* – I'd hope it fares much better than in the wider countryside – but none of them gives us that information for their land-holding.

Praise them as well as being a critical friend. When the magazine or newsletter you receive has news of impressive projects which make a difference for wildlife then point out that you have read it and are impressed by how your money is being spent – you might even enclose a small cheque as a top-up, but do tell them why you gave them that extra cash.

Say that you are reviewing your will this year and have noticed just how many wildlife conservation organisations are doing good things, and that you'd like to make sure any legacies you leave are maximally effective for wildlife.

I'd ask you to contact the organisations you have invested in in this way at least four times a year, and if you support/subscribe to several organisations then I'd ask you to do that for each of them.

And I wouldn't ask you to do this if I weren't prepared to do the same myself, so we can do it together.

Reflection 6

When I am in a hopeful mood I think that human progress might be seen as a process of very gradual readjustment from a selfish society to one which pays more attention to incorporating the principles of fairness. Not just fairness to other members of our own species but also a more sympathetic approach to the other species – the wildlife – which make this planet so special.

The decline in wildlife and the improvement in social equity have both been the products of societal decisions, maybe not very clear decisions but decisions with consequences. If we care about wildlife declines then we have to change the way that our society makes decisions. That's very difficult, and you and I would be hard-pressed to make much difference on our own, but luckily there are purpose-built organisations, wildlife conservation organisations, which can make much more progress with government if they really put their minds to it and if we give them our support. At the moment, too much of that financial support is given through habit, and we need to be more demanding as investors.

The primary players in a wildlife renaissance will have to be government, because government spending, policies and regulations define the context of all activities that affect wildlife. At the moment government does very little to stem wildlife losses – their hearts and heads really aren't in it. We need government to do more, spend more and intervene more to prevent wildlife loss. In theory, in a democracy, that shouldn't be beyond us, but it means harnessing more people to speak up in favour of change that benefits wildlife. This change is not easy to bring about, but it is necessary.

The secondary players in wildlife recovery are wildlife conservation organisations, mostly charities, which we the public support. They are only secondary in the sense that their own power to do good is less than that of government, but their role is crucial. If we had benign wildlife-friendly governments in place then we'd hardly need our wildlife organisations – they could disband or simply plan the next tea shop or sleepover for children. But as we stand at the moment, the role of those conservation organisations is very important because they deliver much of the most effective conservation action, and it is that action which prevents things getting even worse. But also, they are our best way to influence governments. They can reach parts of government that you and I cannot reach, and that's what we should expect those organisations to do for us, in return for our financial and moral support.

The lowly players in wildlife recovery are ourselves as individuals. Despite what you might read elsewhere, you and I are pretty powerless on our own. We have some power and influence and we should be good consumers by exercising choice, and importantly restraint, in what we buy – but the links between what I buy and what wildlife I will see around me are very weak and very uncertain. Don't beat yourself up that you can't save the Turtle Dove by buying a different brand of bread, because you almost certainly can't. Where we can make most difference is in being active wildlife investors through supporting the conservation organisations that make the most difference. So it isn't the bread you buy that will make a difference, but the conservation programme that you support. And you and I need to be critical friends, engaged with the work of our chosen wildlife NGOs. To achieve a top-down influence on wildlife from government we need a grassroots, bottom-up, polite and thoughtful uprising of the millions (not many millions) of supporters of conservation organisations in order to encourage those organisations to become even more active and much more successful in influencing government.

Recapitulation

Wildlife is all around us. In our towns, it is finding cracks in pavements in which to germinate, seeking opportunities to nest in and around our buildings, and entering our houses to scurry around our kitchens, leaving mouse droppings or sitting in the bath looking scary or odd, depending on how you look at it. In our intensively farmed countryside, it is struggling to make a living, to survive in patches of natural habitat, and to prosper in places set aside for wildlife – but it is still there. Every day, as we go about our everyday business, we can all interact with wildlife.

People look at wildlife in very different ways. I love it, I'm fascinated by it, and I have spent much of my life trying to get it a better deal, but not everyone feels like that. For some, the native plants that I spend time enjoying in my rather wild garden are weeds; for some, the birds I enjoy watching are scary feathery things, or even pests. But for many – very many, I believe – wildlife is of only passing interest and little relevance.

Wildlife has taken a mighty clobbering in our country over many centuries, but much of that damage was done a long time ago and the chances of us putting everything back as it was are negligible – and that wouldn't even be the best thing to do.

We can set a very progressive and positive agenda for wildlife in the UK but it is going to have to fit in with 68 million people with a high average standard of living and high expectations of those standards rising. The UK is in the top 30 countries if ranked by per capita Gross Domestic Product, and so dealing with poverty in the UK is a matter of dealing with inequality, not a matter of overall financial poverty. How do we create a wildlife renaissance while living in the real world of our fellow Britons, and on a planet with over a hundred times as many human inhabitants as in our so-called united kingdom?

Those of us who are on wildlife's side, and there are fewer of us truly committed to this cause than everyone thinks, need to realise that we are few. We are few, but we are not powerless. We have some rather limited individual power to influence the future of wildlife in the UK and beyond, but we have much greater collective power. But to harness our power we need to think about our actions, focus our energies on the most important things, and gather together to act cooperatively.

Historically, the British have done their wildlife conservation at arm's length through being members of large (and some small) organisations which collect our money and do good things with it. That is still the best model for action, but it is one that needs refreshing. Our wildlife NGOs have grown stale and are now less effective than they once were. We are the main funders of wildlife NGOs and so we have a lot of unused power in getting them to perform better. We should use that power.

We should also use our political muscle to engineer change and a better future for wildlife. Governments across the UK need to be playing a much bigger role in wildlife conservation – they should own and manage more land, spend more money, and intervene with more regulation in order for wildlife to prosper. Our wildlife NGOs should be helping us to exert pressure on government far more than they are at the moment – so this is an area where we should tell the NGOs we fund to improve their performance. Only when hundreds of thousands of people are visibly mobilised to speak up for wildlife will UK wildlife have a voice at the table where decisions are made. That is what wildlife needs, and that is what our wildlife conservation organisations have to deliver.

There is no doubt that wildlife in the UK is in long-term and medium-term decline, but in the short term it's more difficult to know. There are hopeful signs and we might, we just might, be at a time when things are starting to get better. That should be our aim – to give wildlife a better future in the UK and at the same time to do that in a way that fits in with the rather comfortable and pampered lives that many of us lead. That strikes me as entirely possible to achieve in the lifetime of my children and my grandson, although I think I might only see signs of it happening if I am lucky. That's fine.

There is plenty of hope for the future, but the hope is predicated on action, and the action must be taken by us, not by some nebulous 'them' – and the 'us' needs *you* to play your part to the full. I hope that this book might help you play your part more effectively, but it isn't a complete blueprint for success. If you wait for that you'll be waiting for ever. Engage your brain and then engage your muscles and get things done. Thank you.

Notes, references and further reading

This is a book of ideas rather than a book of facts. So I toyed with the idea of not having any references at all, but decided that would be wrong. In the following notes you will therefore find references of several types: those that establish where I got a particular number or figure, those where I think the reader of this book might want to learn some more, and, just occasionally, those which seemed interesting to me even if they aren't essential. After all, this book is supposed to be interesting and lead you into more and more interesting places.

For example, I use the figure of 70,000 for the number of species in the UK (first in Chapter 2 but a few times afterwards), and in the notes for Chapter 2 you'll find a link to the Natural History Museum website which says that. Now I have no idea whether that is right or not so I'm trusting the NHM to know their job – for that surely is one of their jobs (and it isn't one of mine). I'm trusting the NHM but I'm not endorsing their figure – I'm just telling you where I got it from. And it is probably the case that for the purposes of this book any figure between 7,000 and 700,000 would not materially alter the sense of what I am writing.

Wikipedia is a much-maligned but incredibly useful source of information. I have included only a few Wikipedia references here, where they seemed particularly useful, as you can find them easily yourself.

Since many of the references given here are links to the internet I have added this reference and note section to my website (markavery. info), which will mean that if you have a paper copy of this book in your hand there is a handy place to find a clickable link to all the online references.

Chapter 1: Glimpses of wildlife

Herb-Robert

The RHS has two goes at telling you about Herb-Robert on its website. One (www.rhs.org.uk/plants/7925/geranium-robertianum/details) treats it as a plant you might want to encourage, whereas the other (www.rhs.org.uk/weeds/herb-robert) considers it a problem species.

The Woodland Trust on Herb-Robert: www.woodlandtrust.org.uk/trees-woods-and-wildlife/plants/wild-flowers/herb-robert

Raunds as described by Wikipedia: en.wikipedia.org/wiki/Raunds – and by Raunds Town Council: www.raunds-tc.gov.uk

The Raunds Fish Bar: www.facebook.com/raundsfishbar1

The quiet wood

Glapthorn Cow Pasture is the wood, a wildlife reserve of the Bedfordshire, Cambridgeshire and Northamptonshire Wildlife Trust (www.wildlifebcn.org/nature-reserves/glapthorn-cow-pastures). It is the home of Black Hairstreak butterflies as well as, once, Nightingales. In 2022 I visited at what would normally be peak Nightingale time and there were none, but the Song Thrushes sounded wonderful, and one even had a hint of Nightingale in its song – could it possibly be that a snatch of Nightingale song has been handed down through the local Song Thrushes?

Red Kites

Ian Carter is an expert on this species, and I recommend his recent book. Carter, I. & Powell, D. 2019. *The Red Kite's Year*. Pelagic Publishing, Exeter. pelagicpublishing.com/products/red-kites-year-carter-powell-9781784272005

The Wikipedia entry on shifting baselines is pretty good and explains that an influential use of the concept in ecology was by Daniel Pauly in the context of fisheries management. A healthy baseline ecosystem was probably in place long before we were born, and we shouldn't let our own, limited, lived experience influence our thinking too much. en.wikipedia.org/wiki/shifting_baseline

Hedgehogs

Hugh Warwick is a Hedgehog enthusiast, and his website is a good place to start to find out more about these interesting beasts: www.hughwarwick.com

Jackson, D. 2001. Experimental removal of introduced hedgehogs improves wader nest success in the Western Isles, Scotland. *Journal of Applied Ecology* 38: 802–812. doi.org/10.1046/j.1365-2664.2001.00632.x

Potter, B. 1905. *The Tale of Mrs Tiggy-Winkle*. Frederic Warne and Co., London.

Grey Squirrels

The Woodland Trust has an accurate and sympathetic account of Grey Squirrels here: www.woodlandtrust.org.uk/trees-woods-and-wildlife/animals/mammals/grey-squirrel

Marc Baldwin's Wildlife Online website. Squirrels and forestry: www.
wildlifeonline.me.uk/animals/article/squirrel-interaction-with-humans-
damage-to-forestry

RoSPA website. Tufty. www.rospa.com/about/history/tufty

My former boss at the RSPB, Graham Wynne, pointed out to me that many
people love Grey Squirrels and suggested that I looked at them in the
London Parks. I did. He was right (not for the first time).

Cats

The Conversation website. Pets: is it ethical to keep them? theconversation.com/
pets-is-it-ethical-to-keep-them-115647

Brown Bears in Britain: www.bbc.co.uk/news/science-environment-44699233

Wolves in Britain: en.wikipedia.org/wiki/Wolves_in_Great_Britain

Lynx in Britain: www.bbc.co.uk/news/science-environment-31813207

Falcons and flycatchers

Ratcliffe, D.A. 1980. *The Peregrine Falcon*. T. & A.D. Poyser, Berkhamsted. This is
a good overview of the Peregrine, but I know of no monograph written on
Spotted Flycatchers. Instead, this account on the BTO website is a good introduction to the species: www.bto.org/understanding-birds/species-focus/
spotted-flycatcher

The population trend of Spotted Flycatcher can be seen on the information-
packed but slightly unwieldy Pan-European Common Bird Monitoring
Scheme website: pecbms.info/trends-and-indicators/species-trends/species/
muscicapa-striata. The >40% decline in numbers since 1980 is described as
moderate, as it is less than 5% per annum! Imagine any measure of human
wellbeing that fell by more than half in that period being described as
moderate.

St Peter's Church, Raunds: 4spires.org/church-histories/st-peters-church-
raunds.php

Gells Garage: www.gells-raunds.co.uk

Achurch churchyard no longer seems to have Spotted Flycatchers but is the
last resting place of the Fourth Baron Lilford (www.lilfordhall.com/4th-
baron-lilford.asp), who lived nearby and wrote the most recent county
avifauna of Northamptonshire – published in 1895! www.banburyornitho-
logicalsociety.org.uk/index.php/birds/publications/historic-publications/
birds-of-northamptonshire-lord-lilford

Birds and 'weeds'

The BTO/JNCC/RSPB Breeding Bird Survey homepage gives the rationale of the
study and access to the rather excellent annual reports from it: www.bto.
org/our-science/projects/breeding-bird-survey

Plantlife on arable plants: www.plantlife.org.uk/uk/discover-wild-plants-nature/
habitats/arable-farmland

Black Grass, Black Twitch, Hungerweed, Rat-tail Grass, Slender Foxtail or
Alopecurus myosuroides and its impacts on crops explained, with suggestions

on control from an organic perspective: www.gardenorganic.org.uk/weeds/
   black-grass

Big Garden Birdwatch

The results of the 2022 Big Garden Birdwatch: www.rspb.org.uk/get-involved/
   activities/birdwatch – and some background to the event: www.rspb.org.uk/
   about-the-rspb/about-us/media-centre/press-releases/bgbw-2022-results
Does garden bird feeding do more harm than good? www.bbc.co.uk/news/
   science-environment-58346043
What the BTO says about trichomonosis: www.bto.org/our-science/projects/
   gbw/gardens-wildlife/garden-birds/disease/trichomonosis

House Mice

My guess would be that House Mice are rarer now than they were in my youth
   but I can't find much evidence either way. The Wikipedia entry on House
   Mice told me lots of things about them that I didn't know: en.wikipedia.org/
   wiki/house_mouse
If it is true that the world will beat a path to your door if you build a better mousetrap
   (quoteinvestigator.com/2015/03/24/mousetrap) then I recommend the
   Little Nipper model – I found it fatally effective: www.toolstation.com/
   pest-stop-little-nipper/p46451

Stanwick Lakes

The Stanwick Lakes website (www.stanwicklakes.org.uk) makes it perfectly clear
   that the site is a visitor attraction. On a sunny day, the car park may be
   full to bursting and the queue to get in extends back onto the A45. But
   that is something I rarely see as most of my visits are early in the morning,
   sometimes before the site's car park is even open (by way of a handy layby
   further down the road). Stanwick Lakes is a Site of Special Scientific Interest
   and also a Special Protection Area for birds designated by the UK government
   under our commitments to the EU Birds Directive: www.northampton.gov.
   uk/info/200205/planning-for-the-future/2105/upper-nene-valley-gravel-
   pits-spa-spd and en.wikipedia.org/wiki/Upper_Nene_Valley_Gravel_Pits
An early blog of mine on Turtle Doves: community.rspb.org.uk/ourwork/b/
   markavery/posts/no-turtle-doves-in-any-trees-at-my-place
UK Turtle Dove decline figures: www.bto.org/our-science/projects/bbs/
   latest-results/trend-graphs
European Turtle Dove trend: pecbms.info/trends-and-indicators/species-trends/
   species/streptopelia-turtur
Two local birders, Steve Fisher and Bob Webster, are mates and know this site
   from their youths – they are great birders and interesting companions at
   this site.

Swifts

Gerald Collini spotted the opportunity of refurbishment of the chapel to include
   the addition of a Swift nest box. 2022 seemed quite a good year for Swifts

and I spent some time watching them going in and out of their nest holes on the south side of the chapel one June day. At that stage none was showing any interest in the new nest box provided for them but we'll have to see what happens.

Raunds Methodist chapel: www.rushdenheritage.co.uk/Villages/Raunds/raunds-methodist.html

Ada Salter: www.quakersintheworld.org/quakers-in-action/297/Ada-Salter

David Frost: BBC News. Sir David Frost, broadcaster and writer, dies at 74: www.bbc.co.uk/news/entertainment-arts-23920336

Lack, D. 1956. *Swifts in a Tower*. Methuen, London, was republished in 2018 by Unicorn Press, with a new chapter by Andrew Lack and a beautiful cover by my late colleague Colin Wilkinson; reviewed by me here: markavery.info/2018/05/13/book-review-swifts-in-a-tower-by-david-lack. See also Foster, C. 2021. *The Screaming Sky*. Little Toller Books, Beaminster; reviewed by me here: markavery.info/2021/06/27/sunday-book-review-the-screaming-sky-by-charles-foster

Action for Swifts website: actionforswifts.blogspot.com

Flying Ant Day

The story behind the biology of Flying Ant Day: Hart, A.G., Hesselberg, T., Nesbit, R. & Goodenough, A.E. 2017. The spatial distribution and environmental triggers of ant mating flights: using citizen-science data to reveal national patterns. *Ecography* 41: 877–888. onlinelibrary.wiley.com/doi/epdf/10.1111/ecog.03140

Two examples of headlines. *Daily Mirror*: Flying Ant Day 2020: When is it? What is it? How to get rid of ants in the UK (www.mirror.co.uk/science/what-flying-ant-day-2019-10742632). *Wales Online*: Flying ant day. What is it and how to get rid of them (www.walesonline.co.uk/lifestyle/fun-stuff/flying-ant-day-when-2020-18584649) (note the term 'fun-stuff' in the link).

A good book on ants in general: Jones, R. 2022. *Ants: the Ultimate Social Animals*. Bloomsbury, London; reviewed by me here: markavery.info/2022/01/23/sunday-book-review-ants-by-richard-jones

Blackberries

Plantlife's Seven fascinating facts about blackberries: plantlife.love-wildflowers.org.uk/about_us/blog/blackberry_facts

Apomixis in plants: en.wikipedia.org/wiki/apomixis

Batology – the study of brambles: www.bettsecology.co.uk/insight/batology

Blackberry and apple pie: www.bbcgoodfood.com/recipes/bramley-blackberry-pie

Pheasants

Avery, M.I. 2019. The Common Pheasant: its status in the UK and the potential impacts of an abundant non-native. *British Birds* 112: 372–389. Email me (mark@markavery.info) and I'll send you a pdf of the paper.

A review of the impacts of non-native gamebirds on native wildlife – inadequate but a start: wildjustice.org.uk/gamebird-releases/the-defra-ne-basc-gwct-exeter-university-review-of-released-gamebird-impacts

In 2022, releases of Pheasants into the countryside were very much lower than in other years due to a combination of avian flu outbreaks, Covid-19 and Brexit affecting the ability to import eggs and chicks into the UK, as well as a Wild Justice legal victory which led to restrictions on releases: wildjustice.org.uk/gamebird-releases/wild-justice-statement-on-gamebird-licensing

Bathroom wildlife

Plantlife on Ivy: www.plantlife.org.uk/uk/discover-wild-plants-nature/plant-fungi-species/ivy

Woodland Trust on Ivy: www.woodlandtrust.org.uk/trees-woods-and-wildlife/plants/wild-flowers/ivy

Wildlife Trusts on Ivy Bees: www.woodlandtrust.org.uk/trees-woods-and-wildlife/plants/wild-flowers/ivy

The Royal Entomological Society on silverfish: www.insectweek.co.uk/discover-insects/silverfish-firebrats/silverfish

Everybody, from pest control companies to natural history websites, seems to believe there are 650 spider species in Britain (see www.countryfile.com/wildlife/how-to-identify/british-spider-guide-common-species-to-identify-and-where-to-find-them, for example) but it's not totally clear where that figure comes from. But it sounds plausible, doesn't it?

Easter

Barnack Hills and Holes: langdyke.org.uk/welcome-to-langdyke-countyside-trust/lct-es/barnack-hills-and-holes

Plantlife on Pasqueflowers: www.plantlife.org.uk/uk/discover-wild-plants-nature/plant-fungi-species/pasqueflower

John Clare on Pasqueflowers: www.jeremybartlett.co.uk/2022/04/26/pasque-flower-pulsatilla-vulgaris

Daisies

Plantlife on Daisies: www.plantlife.org.uk/uk/discover-wild-plants-nature/plant-fungi-species/daisy

The RHS on Daisies: www.rhs.org.uk/advice/profile?pid=370

Chapter 2: The State of Wildlife in the UK

What counts?

Live Euro exchange rates against a range of currencies including the pound sterling, updated every minute in working hours and accurate to five significant figures: www.exchangerates.org.uk/Euro-EUR-currency-table.html. At 10:32 a.m. on 16 June 2022 you might get 1.16367 euros to the pound.

Live HSBC share price, updated every minute: www.bing.com/search?q=HSBC +share+value+in+real+time and standing at 523.1p at 10:13 am on 16 June 2022 – check it out for yourself to find out whether we should have bought or sold that day.

Adams, D. 1979. *The Hitchhiker's Guide to the Galaxy.* Pan Books, London. This is the first book in the five-part 'trilogy' by Douglas Adams and is, apart from many other things, a masterpiece of sage advice to environmental campaigners on how the world (and the rest of the universe?) works and what environmental sustainability really means.

How Bhutan measures Gross National Happiness: www.dailybhutan.com/ article/how-does-bhutan-measure-gross-national-happiness-gnh

The New Economic Foundation Happy Planet Index (other indices exist) up to 2019: happyplanetindex.org/hpI – which shows Costa Rica at the top of the list and the UK at number 14.

Why Costa Rica tops the list: www.yesmagazine.org/issue/climate-action/2019/ 01/31/why-costa-rica-tops-the-happiness-index

Defra indicators in 2021: assets.publishing.service.gov.uk/government/uploads/ system/uploads/attachment_data/file/1058728/England_biodiversity_ indicators_2021_FINAL_REVISED_version_3.pdf – but in 2022 many of these indicators, especially the wildlife indicators, were summarily dropped.

UK population census summaries: en.wikipedia.org/wiki/Demography_of_the_ United_Kingdom

There are over 70,000 species of animals, plants, fungI and single-celled organisms found in the UK according to the Natural History Museum: www.nhm.ac.uk/our-science/data/uk-species.html

Wildlife in space and time

The UK *State of Nature* report of 2019, produced by a large consortium of organisations, is a good reference source for examples of UK wildlife declines: nbn.org.uk/stateofnature2019

Lake, S., Liley, D., Still, R. & Swash, A. 2020. *Britain's Habitats: a Field Guide to the Wildlife Habitats of Great Britain and Ireland*, 2nd edition. Wildguides/ Princeton University Press, Princeton, NJ. This is a fine overview of British and Irish habitats – see my review of an earlier edition here: markavery. info/2015/03/15/sunday-book-review-britains-habitats-lake-al

The gold standard?

The BTO/JNCC/RSPB Breeding Bird Survey homepage (www.bto.org/ our-science/projects/breeding-bird-survey) gives the rationale of the study and access to the rather excellent annual reports from it.

Information about the CBC which preceded BBS: www.bto.org/our-science/ publications/birdtrends/2019/methods/common-birds-census – and the book of the scheme's findings: Marchant, J.H., Hudson, R., Carter, S.P. & Whittington, P.A. 1990. *Population Trends in British Breeding Birds*. BTO, Tring.

General information on several other bird monitoring schemes from the BTO: www.bto.org/our-science/projects/birdtrack/surveys

The work of the Rare Breeding Bird Panel: www.rbbp.org.uk

Information about bird surveys from RSPB: www.rspb.org.uk/our-work/ conservation/projects/bird-surveys-in-the-uk

UK extinctions

Extinction rates on planet Earth: Smithsonian Institution. Extinction over time. naturalhistory.si.edu/education/teaching-resources/paleontology/ extinction-over-time

List of extinct British and Irish animals: en.wikipedia.org/wiki/List_of_extinct_ animals_of_the_British_Isles – probably rather unreliable in detail.

There are also microspecies or apomictic plant species, which are lineages of asexually reproducing plants. Rich, T. 2020. List of vascular plants endemic to Britain, Ireland and the Channel Islands 2020. *British & Irish Botany* 2: 169–189 (britishandirishbotany.org/index.php/bib/article/view/51/80). Here, Tim Rich lists 15 extinct UK endemic apomictic plants of which 9 are hawkbits (out of 332 endemic UK hawkbit species) and 4 brambles (out of 194 endemic UK species), and there must be a decent chance that some of these apomictic species will be rediscovered in time. The other two extinct (maybe) endemic species are York Groundsel (a hybrid between the native Common Groundsel and the non-native Sicilian Groundsel which 'escaped' from the Oxford Botanic Garden after being brought there from Mount Etna in the very late seventeenth century – see en.wikipedia.org/wiki/ Senecio_eboracensis) and the Interrupted Brome, a species which benefited from sainfoin cultivation and is the subject of a reintroduction project from seed collections: naturebftb.co.uk/2019/08/01/brome-interrupted.

Three formerly rare species found in the UK, all with continental populations, are thought to have gone extinct:
Downy Hemp-nettle: plantatlas.brc.ac.uk/plant/galeopsis-segetum
Lamb-succory: plantatlas.brc.ac.uk/plant/arnoseris-minima
Davall's Sedge: plantatlas.brc.ac.uk/plant/carex-davalliana

Non-natives

The website of the GB Non-native Species Secretariat: www.nonnativespecies. org/non-native-species

Non-native plants in the UK: Thomas, C.D. & Palmer, G. 2015. Non-native plants add to the British flora without negative consequences for native diversity. *Proceedings of the National Academy of Sciences of the USA* 112: 4387–4392. www.pnas.org/content/112/14/4387

Woodland

Accessible map of ancient woodland in England: naturalengland-defra.opendata. arcgis.com/datasets/Defra::ancient-woodland-england/about

Accessible map of ancient woodland in Wales: naturalresources.wales/evidence- and-data/research-and-reports/ancient-woodland-inventory

Not at all accessible data on ancient woodland in Scotland: www.nature.scot/doc/guide-understanding-scottish-ancient-woodland-inventory-awi

International comparisons of forest cover by country: www.forestresearch.gov.uk/tools-and-resources/statistics/forestry-statistics/forestry-statistics-2018/international-forestry-3/forest-cover-international-comparisons

Woodland Trust. 2020. Disappointing planting figures in England still far below government target. www.woodlandtrust.org.uk/press-centre/2020/06/government-planting-figures/

Plantlife. 2012. *Forestry Recommissioned*. www.plantlife.org.uk/uk/our-work/publications/forestry-recommissioned

A JNCC report on impacts of nitrogen deposition: Stevens, C.J., Smart, S.M., Henrys, P. et al. 2011. Collation of evidence of nitrogen impacts on vegetation in relation to UK biodiversity objectives. JNCC Report 447. hub.jncc.gov.uk/assets/9f1ab259-00f1-4080-9039-ec83e4031db1

Impacts of air pollution on lichens and bryophytes, from Air Pollution Information System: www.apis.ac.uk/impacts-air-pollution-lichens-and-bryophytes-mosses-and-liverworts

Woodland bird declines: jncc.gov.uk/our-work/ukbi-c5-birds-of-the-wider-countryside-and-at-sea

Avery, M.I. & Leslie, R. 1990. *Birds and Forestry*. T. & A.D. Poyser, London.

Crick, H.Q.P., Dudley, C., Glue, D.E. & Thomson, D.L. 1997. UK birds are laying eggs earlier. *Nature* 388: 526. www.nature.com/articles/41453

Thackeray, S.J., Sparks, T.H., Frederiksen, M. *et al.* 2010. Trophic level asynchrony in rates of phenological change for marine, freshwater and terrestrial environments. *Global Change Biology* 16, 3304–3313. onlinelibrary.wiley.com/doi/full/10.1111/j.1365-2486.2010.02165.x

New Scientist, 2 February 2022. UK's spring flowers are blooming a month early due to climate change. www.newscientist.com/article/2306602-uks-spring-flowers-are-blooming-a-month-early-due-to-climate-change

Farmland

International wheat total-production statistics by country: en.wikipedia.org/wiki/international_wheat_production_statistics

UK yields/area for many agricultural products (e.g. wheat, barley, oilseed rape, milk) are high compared with other countries and are regarded as being close to the maximum attainable yields under current conditions of climate and soil, and with existing genetic varieties – see for example wheat: www.fao.org/3/Y4252E/y4252e13.htm – but the Food and Agriculture Organization website has other fascinating examples and insights into agricultural productivity.

The standing crop of farmland birds on British farmland has more than halved in less than 50 years: jncc.gov.uk/our-work/ukbi-c5-birds-of-the-wider-countryside-and-at-sea

Donald, P.F., Green, R.E. & Heath, M.F. 2001. Agricultural intensification and the collapse of Europe's farmland bird populations. *Proceedings of the Royal Society B* 268: 25–29. royalsocietypublishing.org/doi/10.1098/rspb.2000.1325. This paper not only shows (Figure 4) that UK farmland bird declines were higher than those of other European countries but also that a country's farmland bird population trends were related to the intensity of its agriculture.

For a simple overview of causes of farmland bird declines, see www.rspb.org.uk/our-work/conservation/conservation-and-sustainability/farming/near-you/farmland-bird-declines

Newton, I. 2017. *Farming and Birds*. William Collins, London; reviewed by me here: markavery.info/2017/08/16/book-review-farming-birds-ian-newton/

Shrubb, M. 2007. *The Lapwing*. T. & A.D. Poyser, London.

Donald, P. 2004. *The Skylark*. T. & A.D. Poyser, London.

We don't know so much about insect changes on farmland. One of the most interesting studies published in recent years was from Germany, not the UK, and was based on nature reserves, not farmland. However, it is widely regarded as being indicative of what is probably happening more widely across habitats and land uses in many Western European countries. The study involved catching insects in a simple trap and bottling the catches for future analysis. The story goes, perhaps apocryphally, that a visiting researcher noticed the rows of insects in bottles on shelves and asked what they were. When told, he asked whether anyone had noticed that the early bottles, which I imagine as being on the top left of the shelves, had many more insects than the more recent collections. It's a good story but the science that came from that observation is here: Hallmann, C.A., Sorg, M., Jongejans, E. *et al.* 2017. More than 75 percent decline over 27 years in total flying insect biomass in protected areas. *PLoS ONE* 12(10): e0185809. journals.plos.org/plosone/article?id=10.1371/journal.pone.0185809. For a more popular though essentially accurate account, see www.theguardian.com/environment/2017/oct/18/warning-of-ecological-armageddon-after-dramatic-plunge-in-insect-numbers

The marine environment

A *Guardian* article from 2010 that has the eye-watering figures in it about UK fish catches which I used here: www.theguardian.com/environment/2010/may/04/fish-stocks-uk-decline – and if we track down the original scientific paper then we find that the newspaper article is pretty much faithful to the findings of the study: Thurstan, R., Brockington, S. & Roberts, C. 2010. The effects of 118 years of industrial fishing on UK bottom trawl fisheries. *Nature Communications* 1: 15. www.nature.com/articles/ncomms1013

A short, but very interesting, parliamentary debate from 1956 on the subject of the Herring industry: api.parliament.uk/historic-hansard/commons/1956/dec/14/herring-industry

The Continuous Plankton Recorder programme: www.cprsurvey.org

Graphs from the Continuous Plankton Recorder data: www.cprsurvey.org/data/data-charts

Fish stocks: www.theguardian.com/environment/2021/jan/22/only-a-third-of-uks-key-fish-populations-are-not-overfished

Bluefin Tuna coming back to UK waters probably because of warming seas: phys.org/news/2019-01-bluefin-tuna-uk.html

Where we sit in the global wildlife crisis

Living Planet index for 2020: www.livingplanetindex.org

Biodiversity Intactness Index: www.nhm.ac.uk/our-science/data/biodiversity-indicators/about-the-biodiversity-intactness-index.html

The UK's place in the Biodiversity Intactness Index: www.rspb.org.uk/globalassets/downloads/about-us/48398rspb-biodivesity-intactness-index-summary-report-v4.pdf

Wikipedia on global extinction crises: en.wikipedia.org/wiki/Extinction_event

The House Martin effect

House Martin population trends according to the BTO: app.bto.org/birdtrends/species.jsp?year=2019&s=houma

The horsemen of the ecological apocalypse

American biologist Jared Diamond coined the term 'ecological horsemen of the apocalypse' in the 1980s and referred to Invasive Species, Overharvesting, Habitat Destruction and Extinction Chains (this being the extinction of one species causing the extinction of others, for example parasites that only live on one host species). See Diamond, J.M. 1984. 'Normal' extinction of isolated populations. In M.H. NiteckI (ed.), *Extinctions*. Chicago University Press, Chicago, IL, pp. 191–246.

The great E.O. Wilson preferred the term HIPPO, standing for Habitat loss, Invasives, Pollution, Population and Overharvesting. See E.O. Wilson Foundation: eowilsonfoundation.org/how-businesses-can-help-make-half-earth-a-reality-introduction

Both Diamond and Wilson were mostly thinking about extinctions, whereas I am thinking here about any big losses of wildlife and I'm happy to stick to four horsemen; Habitat Loss and Degradation, Invasive Species, Pollution (including climate change) and Overharvesting.

Examples of impacts of non-native invasive plants: www.theconversation.com/invasive-plants-have-a-much-bigger-impact-than-we-imagine-82181

Too many notes, Mozart: www.youtube.com/watch?v=H6_eqxh-Qok&t=31s

Losses of UK habitats:

> Lowland heathland, often quoted as 80% loss in 200 years but probably higher than that and with greater losses in recent decades: data.jncc.gov.uk/data/1ae6a1bf-1dca-4602-9b2e-98ec7bbc5bd4/SSSI-Guidelines-4-LowlandHeathland-2018.pdf

> Wildflower meadows are often said to have declined by 97% (sounds very precise to me) since the 1930s (sounds a little less precise); here is what Kew Gardens says: www.kew.org/read-and-watch/meadows-matter

For ancient woodland loss see references in *Woodland* section, above.

Chalk grassland is widely said to have declined by 80% since World War 2; see what the National Trust says: www.nationaltrust.org.uk/features/whats-special-about-chalk-grassland – and also the words of the Downlands Trust: www.downlandstrust.org.uk/chalk-grassland.html

White-clawed Crayfish www.iucnredlist.org/es/species/2430/9438817#threats

Chapter 3: What is wildlife conservation?

This thing called wildlife conservation

Three different takes on wildlife conservation:

National Geographic's words: education.nationalgeographic.org/resource/wildlife-conservation

Why you should study wildlife conservation at the University of Kent: www.kent.ac.uk/courses/undergraduate/30/wildlife-conservation

The words of Wikipedia: en.wikipedia.org/wiki/Wildlife_conservation

The definition of conservation used by the EU Habitats Directive is 'a series of measures required to maintain or restore the natural habitats and the populations of species of wild fauna and flora at a favourable status.' eur-lex.europa.eu/legal-content/EN/TXT/?uri=CELEX:31992L0043

Monbiot, G. 2013. *Feral.* Allen Lane, London; reviewed by me here: markavery.info/2013/07/14/book-review-feral-george-monbiot

Does it matter?

Stevens, W.K. 2000. Lost rivets and threads, ecosystems pulled apart. *New York Times* (4 July 2000). www.nytimes.com/2000/07/04/science/essay-lost-rivets-and-threads-and-ecosystems-pulled-apart.html

Avery, M.I. 2014. *A Message from Martha: the Extinction of the Passenger Pigeon and its Relevance Today.* Bloomsbury, London.

Sustainable development

The concept of sustainable development was crystallised in a report by the World Commission on Environment and Development: Brundtland, G.H. 1987. *Our Common Future.* WCED. sustainabledevelopment.un.org/content/documents/5987our-common-future.pdf

The moribund website of the UK Sustainable Development Commission retains information on what it did and the reports it issued: www.sd-commission.org.uk

Welsh government sustainability duty, 2019: gov.wales/sites/default/files/publications/2019-07/supplementary-report-to-the-uk-review-of-progress-towards-the-sustainable-development-goals-2030_0.pdf

Report of abandonment of Gwent Levels destructive road: www.theguardian.com/uk-news/2019/jun/04/wales-scraps-gwent-levels-m4-relief-road-scheme

The website of the Future Generations Commissioner for Wales: www.futuregenerations.wales

Mark Drakeford's decision letter on the M4 'relief' road: gov.wales/sites/default/files/publications/2019-06/m4-corridor-around-newport-decision-letter.pdf

Wildlife Trusts report, January 2020, on wildlife losses that will derive from HS2: www.wildlifetrusts.org/news/hs2-exorbitant-cost-nature

Wildlife conservation and climate change action

Nature-based solutions according to IUCN: www.iucn.org/theme/nature-based-solutions

Beebee, T. 2018. *Climate Change and British Wildlife*. Bloomsbury, London. See my review here: markavery.info/2018/11/04/sunday-book-review-climate-change-and-british-wildlife-by-trevor-beebee

Prince of Wales, Juniper, T. and Shuckburgh, E. 2017. *Climate Change*. Michael Joseph, London. See my review here: markavery.info/2017/01/26/sunday-book-review-ladybird-book-climate-change-hrh-prince-charles

Wildlife conservation and animal welfare

The IFAW website exemplifies overlap or confusion, depending on how you look at it, between welfare and conservation issues: www.ifaw.org/uk/journal/facts-animal-welfare-conservation

Simmons, A. 2023. *Treated Like Animals*. Pelagic Publishing, Exeter. See my review here: www.markavery.info/2022/09/18/sunday-book-review-treated-like-animals-by-alick-simms

Several blogs by Alick Simmons also provide a useful introduction to some of the issues, and might persuade you to buy his book:

The ethics of animal exploitation 1: markavery.info/2019/06/07/guest-blog-the-ethics-of-animal-exploitation-part-1-by-alick-simmons

The ethics of animal exploitation 2: markavery.info/2019/06/17/guest-blog-the-ethics-of-animal-exploitation-part-2-by-alick-simmons

The ethics of animal exploitation 3: markavery.info/2019/06/24/guest-blog-the-ethics-of-animal-exploitation-part-3-by-alick-simmons

The ethics of animal exploitation 4: markavery.info/2019/07/10/guest-blog-the-ethics-of-animal-exploitation-part-4-by-alick-simmons

Licensed badger killing: ethical considerations: markavery.info/2019/12/16/guest-blog-licensed-badger-killing-ethical-considerations-by-alick-simmons/.

What conservationists do

Avery, M. 2012. *Fighting for Birds: 25 Years in Nature Conservation*. Pelagic Publishing, Exeter. By me, but by any standards a good insight into what conservation is about and how it is done.

Who are the wildlife conservationists?

The remits of the conservation agencies in all four UK nations:

Natural Resources Wales: gov.wales/sites/default/files/publications/2020-05/natural-resources-wales-nrw-remit-letter-2020-to-2021.pdf. For a more digestible but less definitive account: naturalresources.wales/about-us

Department of Agriculture and Environment Northern Ireland: www.
daera-ni.gov.uk/about-daera

Northern Ireland Environment Agency: www.daera-ni.gov.uk/
northern-ireland-environment-agency

NatureScot: www.nature.scot/doc/applying-naturescots-balancing-duties-
guidance-notice

Natural England: www.gov.uk/government/organisations/natural-england/
about

Joint Nature Conservation Committee: jncc.gov.uk/our-role

Friends of the Earth's history in its own words: friendsoftheearth.uk/who-we-are/
our-history

Greenpeace website: www.greenpeace.org.uk

BASC website on wildlife conservation: basc.org.uk/category/conservation

Game and Wildlife Conservation Trust charitable objects: www.gwct.org.uk/
about/our-charitable-objects

BTO charitable objects, from the Charity Commission website (register-of-char-
ities.charitycommission.gov.uk/charity-search/-/charity-details/216652/
governing-document): 'To promote, organise, carry on and encourage
study and research and particularly field work for the advancement of
knowledge in all branches of the science of ornithology. Permanently to
preserve and protect lands and objects which by their natural features are
suitable for the preservation and study of bird life and of fauna and flora
generally.' As far as I can tell, the second sentence means that the BTO has a
small wildlife reserve attached to its headquarters.

Wildlife and Countryside Link: www.wcl.org.uk

Northern Ireland Environment Link: www.nienvironmentlink.org

Scottish Environment Link: www.scotlink.org

Wales Environment Link: www.waleslink.org

Chapter 4: Wildlife Conservation Successes

Agricultural policy

This section is very much based on an article I wrote for *British Birds* and I am
grateful to the journal for permission to reproduce large chunks of it
here. The optimistic last sentence was written before the advent of a Truss
government. Avery, M.I. 2022. Agricultural policy – what does it mean for
our farmland birds? *British Birds* 113: 310–312.

For something as large and complicated as an EU policy which has changed over
many decades, Wikipedia comes into its own in a big way: en.wikipedia.
org/wiki/Common_Agricultural_Policy

Two good, intelligible, accounts of public and private goods: What are public
goods? Definition, how they work, and example: www.investopedia.
com/terms/p/public-good.asp – and Definition of public good: www.
economicshelp.org/micro-economic-essays/marketfailure/public-goods

An account not just of the MacSharry reforms but also of the background
   policy and something of Ray MacSharry himself from an Irish perspective:
   O'Toole, P. 2021. When farming changed forever: the MacSharry reforms
   and why they still matter. *Irish Farmers Journal.* www.farmersjournal.ie/
   when-farming-changed-forever-the-macsharry-reforms-and-why-they-
   still-matter-624966

A thesis which examines the context of CAP reform and its impact on farming in
   one EU region: Stewart, R.S. 2000. The impact of the 1992 MacSharry CAP
   reforms on agriculture in Grampian Region. PhD thesis, Robert Gordon
   University. rgu-repository.worktribe.com/output/248012

What the EU says about the Common Agriculture Policy and its future
   now: The common agricultural policy (CAP) is about food, the environment
   and the countryside. agriculture.ec.europa.eu/common-agricultural-
   policy_en

Ailsa Craig

Zonfrillo,B.2001.AilsaCraig:beforeandaftertheeradicationofratsin1991.*Ayrshire
   Birding.* www.ayrshire-birding.org.uk/2001/01/ailsa_craig_before_and_
   after_the_eradication_of_rats_in_1991

Glasgow Herald, 18 September 2002. Rat killers have brought the puffins back
   to Ailsa Craig. www.heraldscotland.com/news/11963605.rat-killers-have-
   brought-the-puffins-back-to-ailsa-craig-island-project-is-hailed-a-
   success-as-breeding-birds-return

Bernie Zonfrillo carried out this study for his PhD: theses.gla.ac.uk/71773

Flow Country

Avery, M.I. & Leslie, R. 1990. *Birds and Forestry.* T. & A.D. Poyser, London.

The Flow Country: www.theflowcountry.org.uk

Lindsay, R., Charman, D.J., Everingham, F. *et al.* 1988. The Flow Country: the
   peatlands of Caithness and Sutherland. Nature Conservancy Council,
   Peterborough.

London School of Economics, 22 April 2015. The top rate of income tax. blogs.
   lse.ac.uk/politicsandpolicy/the-top-rate-of-income-tax-2

Nigel Lawson's speech in the 1988 budget where he abolished Schedule D tax
   relief: www.margaretthatcher.org/document/111449

NatureScot's Sitelink, which contains information about notified and designated
   areas: sitelink.nature.scot/home

The MAGIC system is a very useful free resource that contains masses of useful
   and interesting mapped information about designations across the UK.
   This is the place to see what piece of land has what wildlife designation. It is
   slightly offputting but once you've played around with it for a while it opens
   up lots of sources of information on ownership, land use, designation etc.
   magic.defra.gov.uk

Fonseca's Seed Fly
Buglife: www.buglife.org.uk/news/coul-links-seed-fly-added-to-global-at-risk-register
Wikipedia: en.wikipedia.org/wiki/Botanophila_fonsecai
Discover Wildlife website: www.discoverwildlife.com/animal-facts/insects-invertebrates/fonsecas-seed-fly
Guest blog by Jonny Hughes: markavery.info/2018/08/09/guest-blog-coul-links-by-jonny-hughes
It seems, as of July 2022, that the developer is coming back to have another go: markavery.info/2022/07/08/press-release-internationally-protected-coul-links-once-again-under-threat-from-golf-course
I am grateful to Matt Shardlow, CEO of Buglife, for bringing the Fonseca's Seed Fly to my attention.

Hedgerows
Hedgerow protection guidance from Natural England and Defra. Countryside hedgerows: protection and management: www.gov.uk/guidance/countryside-hedgerows-regulation-and-management
Hedgelink: hedgelink.org.uk/hedge-hub
Hedgerows: laws, rules and regulations: www.hedgelink.org.uk/cms/cms_content/files/489_hedges_and_the_law_updated_april_2015.pdf
McCollin, D. 2000. Editorial: Hedgerow policy and protection – changing paradigms and the conservation ethic. *Journal of Environmental Management* 60: 3–6.citeseerx.ist.psu.edu/viewdoc/download?doi=10.1.1.477.2540&rep=rep1&type=pdf
Woodlands.co.uk blog: www.woodlands.co.uk/blog/flora-and-fauna/hedgerow-loss
Fiennes, J. 2022. *Land Healer: How Farming Can Save Britain's Countryside.* Witness Books, London. Jake Fiennes' book has a very good first chapter which discusses hedgerows from the land manager's point of view as well as that of the naturalist – reviewed by me here: markavery.info/2022/07/10/sunday-book-review-land-healer-by-jake-fiennes

Hope Farm and Skylark patches
Hope Farm bird increases: www.rspb.org.uk/our-work/conservation/conservation-and-sustainability/farming/hope-farm/bird-numbersandwww.rspb.org.uk/globalassets/downloads/documents/conservation--sustainability/hope-farm/hope-farm---territory-map.pdf
Hope Farm Annual Review, including details of the economics: www.rspb.org.uk/globalassets/downloads/documents/conservation--sustainability/hope-farm/hope-farm-annual-review-2019.pdf
Defra. The guide to cross-compliance in England 2022: assets.publishing.service.gov.uk/media/623b14228fa8f540eacc2f37/Guide_to_cross_compliance_in_England_2022.pdf

Knepp

Read more about the rewilding aspects of the Knepp project here: www.
rewildingbritain.org.uk/rewilding-projects/knepp-castle-estate
> ... and in Isabella Tree's excellent book: Tree, I. 2018. *Wilding*. Thames &
> Hudson, London; reviewed by me here: markavery.info/2018/07/29/
> sunday-book-review-wilding-by-isabella-tree

Buy their meat here: www.kneppwildrangemeat.co.uk
> ... view the accommodation options here: www.kneppsafaris.co.uk
> ... and see the range of costed visiting options here: www.kneppsafaris.
> co.uk/prices

Lead ammunition

Royal Commission on Environmental Pollution. 1983. Ninth Report: Lead in the
Environment (Cmnd 8852). discovery.nationalarchives.gov.uk/details/r/
C303795

Hansard, 18 April 1983, lead in the environment: hansard.parliament.uk/
Lords/1983-04-18/debates/e7382987-7a1d-4730-b057-015a009f2d54/
LeadInTheEnvironment

Lead Ammunition Group website: www.leadammunitiongroup.org.uk. This
group was set up in the dying months of the last Westminster Labour
government in 2010 to review health and environmental impacts of lead
ammunition use, and it produced a massive report for Defra in 2015:
www.leadammunitiongroup.org.uk/reports. The report recommended the
phasing out of lead ammunition but received no government response for
over a year, until under cover of the resignation of David Cameron after
the Brexit referendum the then Secretary of State for the Environment, Liz
Truss, rejected the report on spurious grounds: markavery.info/2016/07/14/
truss-misleads-waterfowl-science

The Oxford Lead Symposium of 2016 (oxfordleadsymposium.info/proceedings)
contains a wide range of intelligible but rigorous articles on the effects of
lead on people and wildlife and what has been done across the world to
reduce its impacts.

Stroud, D.A., Pain, D.J. & Green, R.E. 2021. Evidence of widespread illegal
hunting of waterfowl in England despite partial regulation of the use of lead
shotgun ammunition. *Conservation Evidence Journal* 18: 18–24. conserva-
tionevidencejournal.com/reference/pdf/9454

Wild Justice collected game meat on sale in UK supermarkets (Sainsbury's,
Harrods and Waitrose) in 2021 and 2022 and had their lead levels
analysed. Samples of chicken, pork and domesticated duck had very low
lead levels: 0% were above the maximum level set by regulation. There
are no maximum lead levels set for game meat, even though it has lead
ammunition shot into it. Lead levels in game meat across the three supermarkets were higher than would be legal were maximum lead levels
imposed on game meat as with other meats, in 85% of 117 samples
(pheasant, partridge, woodpigeon). wildjustice.org.uk/lead-ammunition/
more-high-lead-levels-in-waitrose-game-meat

Food Standards Agency advice on eating lead: 'Consuming lead is harmful, health experts advise to minimise lead consumption as much as possible. Anyone who eats lead-shot game should be aware of the risks posed by consuming large amounts of lead, especially children and pregnant women.' www.food.gov.uk/safety-hygiene/lead-shot-game

NHS England advice to pregnant mothers on eating lead: Foods to avoid in pregnancy – What to avoid – game meats such as goose, partridge or pheasant. www.nhs.uk/pregnancy/keeping-well/foods-to-avoid

UK petrol goes lead-free in 1999: www.theguardian.com/uk/2002/aug/15/oil. business

Lead–crime hypothesis: en.wikipedia.org/wiki/Lead%E2%80%93crime_hypothesis

George Monbiot writing in *The Guardian*, 2013: www.theguardian.com/commentisfree/2013/jan/07/violent-crime-lead-poisoning-british-export

I've been involved in the lead ammunition issue for more than 25 years and I am always surprised we haven't made more progress on it. We have made progress, but very slowly. That progress has been made is largely due to a small group of people, among whom Debbie Pain deserves an enormous share of the credit (listen to her interview on BBC Radio 4's *The Life Scientific* here: www.bbc.co.uk/sounds/play/m000j21k). Also of great importance in this story in the UK have been Gwyn Williams, Rhys Green and Rob Sheldon (all former RSPB staff) and Ruth Cromie (former WWT staff). But the two organisations, RSPB and WWT, have been feeble in this debate and I blame them, quite considerably, for the slow progress on an easily winnable issue. Their feebleness would make a good case study in NGO failure. In contrast, John Swift, a former CEO of BASC, has been a force for good in chairing the Lead Ammunition Group and receiving huge amounts of abuse from the shooting community (as has Debbie Pain and to a lesser extent myself).

Marine protected areas

BBC website. 2014. Lundy's marine recovery. www.bbc.co.uk/devon/content/articles/2008/01/25/lundy_marine_reserve_feature.shtml

Davies, C.E., Johnson, A.F., Wootton, E.C. *et al.* 2015. Effects of population density and body size on disease ecology of the European lobster in a temperate marine conservation zone. *ICES Journal of Marine Science* 72: i128–i138. doi.org/10.1093/icesjms/fsu237

Hoskin, M.G., Coleman, R.A., von Carlshausen, E. & Davis, C.M. 2011. Variable population responses by large decapod crustaceans to the establishment of a temperate marine no-take zone. *Canadian Journal of Fisheries and Aquatic Sciences* 68: 185–200. cdnsciencepub.com/doi/abs/10.1139/F10-143

Marine and Coastal Access Act 2009: www.legislation.gov.uk/ukpga/2009/23/contents

The roll-out of Marine Conservation Zones in inshore English, Welsh and Northern Irish waters and offshore UK waters, from the JNCC website: jncc.gov.uk/our-work/marine-conservation-zones

The Benyon report on Highly Protected Marine Protection Areas: assets. publishing.service.gov.uk/government/uploads/system/uploads/attachment_data/file/890484/hpma-review-final-report.pdf

The Marine Management Organisation website: www.gov.uk/government/organisations/marine-management-organisation

No Mow May

Plantlife:

>www.plantlife.org.uk/uk/discover-wild-plants-nature/no-mow-may

>www.plantlife.org.uk/uk/about-us/news/no-mow-may-how-to-get-ten-times-more-bees-on-your-lockdown-lawn

>www.plantlife.org.uk/uk/about-us/news/how-to-mow-your-lawn-for-wildlife

No Mow May coverage:

>*Gardens Illustrated*: www.gardensillustrated.com/feature/lawn-mowing-when-flowers-may

>*Country Living*: www.countryliving.com/uk/homes-interiors/gardens/a36325312/no-mow-may-plantlife

>A lawn 'care' company: www.lawn-tech.co.uk/blog/home-page/should-i-take-part-in-no-mow-may

No Mow May taken up by Appleton, Wisconsin, with an interesting take on the challenges and opportunities in the USA: beecityusa.org/no-mow-may

Otters

Press coverage of the return of Otters to every English county announced in 2011:

>*Independent*: www.independent.co.uk/climate-change/news/otters-return-to-every-county-in-england-2339626.html

>*Guardian*: www.theguardian.com/environment/2011/aug/18/otters-return-british-rivers

Chanin, P.R.F. & Jefferies, D.J. 1978. The decline of the otter *Lutra lutra* L. in Britain: an analysis of hunting records and discussion of causes. *Biological Journal of the Linnean Society* 10: 305–328. doi.org/10.1111/j.1095-8312.1978.tb00018.x

MacDonald, S.M. 1983. The status of the otter (*Lutra lutra*) in the British Isles. *Mammal Review* 13: 11–23. doi.org/10.1111/j.1365-2907.1983.tb00260.x

Environment Agency. 2010. Fifth otter survey of England 2009–2010: technical report. ptes.org/wp-content/uploads/2015/06/National-Otter-Survey.pdf

Chanin, P. 2003. *Ecology of the European Otter*. Conserving Natura 2000 Rivers Ecology Series No. 10. English Nature, Peterborough. publications.naturalengland.org.uk/publication/81053

Carrell, S. Mink numbers drop as the otter bites back. *The Independent*, 26 October 2003. www.independent.co.uk/climate-change/news/mink-numbers-drop-as-the-otter-bites-back-93043.html

Maxwell, G. 1960. *Ring of Bright Water*. Longmans, Green & Co, London. Details of the 1969 film here: www.imdb.com/title/tt0064893/fullcredits

Williamson, H. 1927. *Tarka the Otter: His Joyful Water-Life and Death in the Country of Two Rivers.* Putnam, London. www.henrywilliamson.co.uk/ bibliography/a-lifes-work/tarka-the-otter/57-uncategorised/148-tarka-the-otter-uk-editions – and the 1979 film: https://www.henrywilliamson. co.uk/bibliography/a-lifes-work/tarka-the-otter/57-uncategorised/150-tarka-the-otter-the-film-and-the-opera

A video of otter hounds in Wiltshire from 1921: www.youtube.com/ watch?v=I17-7e2mTlk

Peregrines and pesticides

Greenwood, J. 2001. It was not DDT. *British Birds* 114: 248–250. britishbirds. co.uk/content/it-was-not-ddt

Cramp, S. 1963. Toxic chemicals and birds of prey. *British Birds* 56: 124–139.

Cyclodienes, from the *Encyclopedia of Toxicology*, 3rd edition, 2014: www. sciencedirect.com/science/article/pii/B9780123864543001184?via%3Dihub

Stone, A. 2019. How saving ozone layer in 1987 slowed global warming: phys. org/news/2019-12-ozone-layer-global.html

Doniger, D. 2019. We saved the ozone layer, we can save the climate: www.nrdc. org/experts/david-doniger/we-saved-ozone-layer-we-can-save-climate

British Antarctic Survey. The ozone hole: www.bas.ac.uk/about/antarctica/ geography/ozone

Hayman, G.D. 1997. CFCs and the ozone layer. British Journal of Clinical Practice, Supplement 89: 2–9. pubmed.ncbi.nlm.nih.gov/9519506

Indian vulture crisis due to diclofenac, vulture safety zones: www.pmfias.com/ diclofenac-indian-vulture-crisis

Buglife. Neonicotinoid insecticides: www.buglife.org.uk/campaigns/pesticides/ neonicotinoid-insecticides

Carson, R. 1962. *Silent Spring.* Houghton Mifflin, New York.

Pine Martens

The Vincent Wildlife Trust project in Wales: www.vwt.org.uk/projects-all/pine-marten-recovery-project – and news of its success: www.bbc.co.uk/news/ uk-wales-54060331

The Gloucestershire Wildlife Trust project in the Forest of Dean: www.glouces-tershirewildlifetrust.co.uk/project-pine-marten

The convincing scientific paper demonstrating the impact of Pine Martens on Grey Squirrels (negative) and Red Squirrels (positive): Sheehy, E., Sutherland, C., O'Reilly, C. & Lambin, X. 2018. The enemy of my enemy is my friend: native pine marten recovery reverses the decline of the red squirrel by suppressing grey squirrel populations. *Proceedings of the Royal Society B* 285: 20172603. doi.org/10.1098/rspb.2017.2603 – and a popular account of that work: www.britishredsquirrel.org/grey-squirrels/pine-martin

Monbiot, G. How to eradicate grey squirrels without firing a shot. *The Guardian*, 30 January 2015. www.theguardian.com/environment/2015/jan/30/how-to-eradicate-grey-squirrels-without-firing-a-shot-pine-martens

Protection of Birds Act 1954

Moving second reading of the PBA: hansard.parliament.uk/Commons/1953-
12-04/debates/fdbb3f8e-77c8-4864-ac94-abc8268328af/ProtectionOfBirds
Bill?highlight=protection%20birds#main-content

UK Parliament. What was the legacy of the Protection of Birds Act (1954)? www.
parliament.uk/about/living-heritage/transformingsociety/towncountry/
landscape/case-study-tufton-beamish-and-wildlife-conservation/
life-and-career/early-life111

Lady Tweedsmuir: en.wikipedia.org/wiki/Priscilla_Buchan,_Baroness_Tweedsmuir_
of_Belhelvie

Tufton Beamish: en.wikipedia.org/wiki/Tufton_Beamish,_Baron_Chelwood

Lists of successful Private Members' Bills in UK parliament: www.parliament.
uk/globalassets/documents/commons-information-office/l03.pdf and
commonslibrary.parliament.uk/research-briefings/sn04568

Reform or abolition of driven grouse shooting

This is a subject close to my heart and in which I have been much involved.
I even wrote a book about it: Avery, M.I. 2016. *Inglorious: Conflict in the
Uplands*, 2nd edition. Bloomsbury, London.

Much of the kicking off of debate on this issue was done by myself (I think I can
claim) in concert with two mates, Chris Packham and Ruth Tingay. Chris
is well known to many from his TV appearances and Ruth is well known to
a smaller more select group of people because of her excellent blog, Raptor
Persecution UK (raptorpersecutionuk.org).In autumn 2018, after Chris's
Walk for Wildlife march in London, the three of us formed the not-for-
profit organisation, Wild Justice (wildjustice.org.uk), which has taken legal
action against public bodies on behalf of wildlife and which campaigns for
a better wildlife future.

But nowadays much of the progress is being made in Scotland, where I don't
live, by a coalition of organisations called Revive (revive.scot). It's a pretty
good case study of how progress can be made more quickly in one part of
the UK, because of a more favourable political landscape (Scotland under
an SNP government), than in another (England under a Tory government).
In England, most of the best advocacy in recent years has been done by
Wild Moors (www.wildmoors.org.uk), led by Luke Steele. This is another
area where the RSPB and the Wildlife Trusts have been feeble. Neither organisation is a member of the Revive coalition. The RSPB did a lot of the
early work on highlighting the problems of moorland burning but has not
been anything like as active as it should have been, and the Wildlife Trusts
have been even more of a house divided on this issue than on many others.
The considerable progress which has been made has been despite the poor
showing by our two premier wildlife conservation organisations.

An important scientific analysis of the fate of satellite-tagged Hen Harriers
confirmed that illegal persecution of protected raptors on grouse moors is
rife, and at high enough levels to severely limit their populations. The paper

was published in *Nature Communications* (doi.org/10.1038/s41467-019-09044-w), but this interpretation of it on my blog may be more immediately intelligible to the non-specialised reader markavery.info/2019/03/19/long-awaited-scientific-paper-nails-grouse-moor-crimes

Climate Change Committee. 2021. Progress in adapting to climate change. Report to Parliament. www.theccc.org.uk/wp-content/uploads/2021/06/Progress-in-adapting-to-climate-change-2021-Report-to-Parliament.pdf – see page 62 for the words 'all rotational burning [of blanket bogs] should cease immediately'.

A local campaign in Hebden Bridge seeking to ban upland burning for grouse shooting upstream of their town: www.hebdenbridge.co.uk/news/2014/045.html

Bradford Council ends grouse shooting on Ilkley Moor: markavery.info/2018/01/16/35804

Yorkshire Post, 30 July 2021. Yorkshire Water to end grouse shooting tenancies on two of its moors with eight more up for review. www.yorkshirepost.co.uk/news/people/yorkshire-water-to-end-grouse-shooting-tenancies-on-two-of-its-moors-with-eight-more-up-for-review-3328082

Sites of Special Scientific Interest

Areas of Special Scientific Interest (ASSIs) are the close equivalents of SSSIs in Northern Ireland and the Isle of Man.

The original National Parks and Access to the Countryside Act 1949: www.legislation.gov.uk/ukpga/Geo6/12-13-14/97/enacted

The Lawton Review: www.gov.uk/government/news/making-space-for-nature-a-review-of-englands-wildlife-sites-published-today

SSSIs and ASSIs:

   Wikipedia: en.wikipedia.org/wiki/Site_of_Special_Scientific_Interest
   JNCC: jncc.gov.uk/our-work/guidelines-for-selection-of-sssis
   Natural England: adlib.everysite.co.uk/resources/000/059/648/NE54.pdf
   NatureScot: www.nature.scot/professional-advice/protected-areas-and-species/protected-areas/national-designations/sites-special-scientific-interest-sssis
   Department of Agriculture, Environment and Rural Affairs (Northern Ireland): www.daera-ni.gov.uk/topics/land-and-landscapes/areas-special-scientific-interest
   Natural Resources Wales: naturalresources.wales/guidance-and-advice/environmental-topics/wildlife-and-biodiversity/protected-areas-of-land-and-seas/sites-of-special-scientific-interest-responsibilities-of-owners-and-occupiers

The Habitats Directive in a nutshell: ec.europa.eu/environment/nature/legislation/habitatsdirective

The Birds Directive in a nutshell: ec.europa.eu/environment/nature/legislation/birdsdirective

Condition assessment figures for English SSSIs: assets.publishing.service.gov.uk/government/uploads/system/uploads/attachment_data/file/925414/1_Extent__and_condition_of_protected_areas_2020_accessible.pdf

Stone-curlews

Considering that the RSPB has done a fantastic job in being the driving force behind Stone-curlew conservation, it is surprisingly difficult to find anything other than a fairly glib, out-of-date account of that work on its website – here is that account: www.rspb.org.uk/birds-and-wildlife/wild-life-guides/bird-a-z/stone-curlew/conservation. I found it easier to get a good overview through this Back from the Brink webpage, which took me to an account of the RSPB work in Wessex: naturebftb.co.uk/2019/05/21/protecting-the-stone-curlew-in-wessex

Information for farmers on how to create Lapwing and Stone-curlew nesting plots: defrafarming.blog.gov.uk/create-nesting-plots-for-lapwing-and-stone-curlew

Betton, K. Saving the stone-curlew. *BirdGuides*, 24 April 2022. www.birdguides.com/articles/conservation/saving-the-stone-curlew

Former RSPB scientist, Professor Rhys Green, did a fantastic job in studying Stone-curlews to elucidate what was needed for their conservation. His findings (working with others) were the foundation for a Stone-curlew recovery plan which was very successful.

Wildlife reserves

Land area managed by the National Trust: en.wikipedia.org/wiki/National_Trust

Land area managed by the National Trust for Scotland: en.wikipedia.org/wiki/National_Trust_for_Scotland

Land area managed by the RSPB: www.countrylife.co.uk/country-life/who-owns-britain-top-uk-landowners-20178

Land area managed by the Wildlife Trusts: www.wildlifetrusts.org/farming

More than 18,500 species are found on RSPB wildlife reserves: www.rspb.org.uk/about-the-rspb/about-us/media-centre/press-releases/2020-an-amazing-year-for-wildlife-on-rspb-reserves

Chapter 5: Why are we failing so badly?

Why might we be failing?

I found this article on climate-change excuses quite interesting: www.vox.com/energy-and-environment/2019/5/17/18626825/alexandria-ocasio-cortez-greta-thunberg-climate-change

Lack of government investment

UK government ministers: www.gov.uk/government/ministers

UK government spending: en.wikipedia.org/wiki/Government_spending_in_the_United_Kingdom

Budget 2021: https://assets.publishing.service.gov.uk/government/uploads/system/uploads/attachment_data/file/966868/BUDGET_2021_-_web.pdf

Defra report and accounts 2020–21: assets.publishing.service.gov.uk/
government/uploads/system/uploads/attachment_data/file/1037320/
defra-year-end-accounts-2020-2021.pdf. It would be very difficult to pin
down the money spent on wildlife conservation from this account.

The unused powers of governments

McCarthy, D. & Morling, P. 2015. *Using Regulation as a Last Resort: Assessing
the Performance of Voluntary Approaches*. Royal Society for the Protection
of Birds: Sandy. Unfortunately this report has disappeared from the RSPB
website but an account of it can be found here: tabledebates.org/research-
library/using-regulation-last-resort-assessing-performance-voluntary-
approaches – and if you'd like a copy of the pdf then email me at mark@
markavery.info and I'll send you one (I'm sure the RSPB won't mind).

Vested interests

National Farmers' Union: www.nfuonline.com
International Air Transport Association: www.iata.org/en/about
National Federation of Builders: www.builders.org.uk/the-nfb
British Association for Shooting and Conservation: basc.org.uk
CropLife UK: www.croplife.uk
UK Petroleum Industry Association: www.ukpia.com
BBC reports over 500 fossil fuel lobbyists at climate change talks in Glasgow
www.bbc.co.uk/news/science-environment-59199484

Wildlife conservation is not seen as a political issue

Conservative general election manifestoes:
    2019: www.conservatives.com/our-plan/conservative-party-manifesto-2019
    2017: general-election-2010.co.uk/conservative-manifesto-2017-pdf-download
    2015: ucrel.lancs.ac.uk/wmatrix/ukmanifestos2015/localpdf/Conservatives.
      pdf
    2010: conservativehome.blogs.com/files/conservative-manifesto-2010.pdf
Labour general election manifestoes:
    2019: labour.org.uk/manifesto-2019
    2017: labour.org.uk/manifesto-2017
    2015:labourlist.org/2015/04/britain-can-do-better-read-labours-2015-
      manifesto-in-full
    2010: www.cpa.org.uk/cpa_documents/TheLabourPartyManifesto-2010.pdf
Liberal Democrat general election manifestoes:
    2019: www.libdems.org.uk/plan
    2017: general-election-2010.co.uk/liberal-democrats-manifesto-2017-pdf-
      download
    2015: d3n8a8pro7vhmx.cloudfront.net/libdems/pages/8907/attachments/
      original/1429028133/Liberal_Democrat_General_Election_
      Manifesto_2015.pdf
    2010: general-election-2010.co.uk/2010-general-election-manifestos/Liberal-
      Democrat-Party-Manifesto-2010.pdf

Green Party general election manifestoes:
>2019: www.greenparty.org.uk/assets/files/Elections/Green%20Party%20
Manifesto%202019.pdf
>2017: www.maniffesto.com/documents/green-party-manifesto-2017
>2015: https://www.greenparty.org.uk/we-stand-for/2015-manifesto.html
>2010: general-election-2010.co.uk/green-party-manifesto-2010-general-election/

Scottish National Party Scottish Parliament election manifestoes:
>2021: www.snp.org/manifesto
>2016: www.snp.org/the-snp-2016-manifesto-explained
>2011: vote.snp.org/campaigns/SNP_Manifesto_2011_lowRes.pdf
>2007: www.theguardian.com/politics/2007/apr/12/scotland.devolution1

Plaid Cymru Welsh Senedd/Assembly election manifestoes:
>2021: manifesto.deryn.co.uk/plaid-cymru-let-us-face-the-future-together
>2016: www.maniffesto.com/documents/plaid-cymru-assembly-election-manifesto-2016
>2011: www.maniffesto.com/documents/plaid-cymru-assembly-election-manifesto-2011
>2007: www.maniffesto.com/documents/plaid-cymru-assembly-election-2007

Badger Trust. Badger cull facts – including information on the cost of the Badger cull and the number of Badgers killed: www.badgertrust.org.uk/badger-cull-facts

How can you tell when a politician is lying? When their lips are moving: constitutionstudy.com/2017/08/07/how-can-you-tell-when-a-politician-is-lying

A house divided – a case study

Wildlife Trusts, how we are run: www.wildlifetrusts.org/how-we-are-run

The incomes of the English and Welsh Wildlife Trusts, and the Royal Society of Wildlife Trusts, can be found by means of multiple searches of the Charity Commission website: register-of-charities.charitycommission.gov.uk/charity-search

For the Scottish Wildlife Trust: www.oscr.org.uk/about-charities/search-the-register

For the Ulster Wildlife Trust: www.charitycommissionni.org.uk

For the Manx Wildlife Trust: www.gov.im/charities

For the Alderney Wildlife Trust: www.charity.org.gg

These are the results (all figures in millions for 2020/21):
>England: Avon Wildlife Trust £3.4, Wildlife Trust for Bedfordshire, Cambridgeshire and Northamptonshire £5.4, Berks, Bucks and Oxon Wildlife Trust £7.0, Birmingham and the Black Country Wildlife Trust 1.3, Cheshire Wildlife Trust £2.3, Cornwall Wildlife Trust £4.6, Cumbria Wildlife Trust £4.0, Derbyshire Wildlife Trust 2.9, Devon Wildlife Trust £5.5, Dorset Wildlife Trust £3.5, Durham Wildlife Trust £2.0, Essex Wildlife Trust £10.0, Gloucestershire Wildlife Trust £4.8, Hampshire and Isle of Wight Wildlife Trust £6.7, Herefordshire

Wildlife Trust £1.8, Herts and Middlesex Wildlife Trust £2.2, Isles of Scilly Wildlife Trust £0.3, Kent Wildlife Trust £6.4, Lancashire Wildlife Trust £6.0, Leicestershire and Rutland Wildlife Trust £2.1, Lincolnshire Wildlife Trust 2.8, London Wildlife Trust £4.6, Norfolk Wildlife Trust £7.2, Northumberland Wildlife Trust £3.0, Nottinghamshire Wildlife Trust £6.3, Sheffield and Rotherham Wildlife Trust £2.7, Shropshire Wildlife Trust £2.2, Somerset Wildlife Trust £3.8, Staffordshire Wildlife Trust £5.8, Suffolk Wildlife Trust £7.2, Surrey Wildlife Trust £7.2, Sussex Wildlife Trust £7.1, Tees Valley Wildlife Trust £0.4, Warwickshire Wildlife Trust £8.1, Wiltshire Wildlife Trust £5.9, Worcestershire Wildlife Trust £2.6, Yorkshire Wildlife Trust £11.0.

Wales: Gwent Wildlife Trust £0.9, North Wales Wildlife Trust £3.1, South and West Wales Wildlife Trust £2.7, Radnorshire Wildlife Trust £0.5, Montgomeryshire Wildlife Trust £0.9.

Scotland: Scottish Wildlife Trust: £6.1

Northern Ireland: Ulster Wildlife Trust: £2.5

Non UK: Manx Wildlife Trust: £0.4

Abraham Lincoln's address to the Illinois Republican State Convention, 1858 (the 'House Divided' speech): www.nps.gov/liho/learn/historyculture/housedivided.htm

A house divided writ large – the tangled bank

The tangled bank: Darwin, C. 1859. *On the Origin of Species*. John Murray, London. darwin-online.org.uk/content/frameset?itemID=F373&viewtype =text&pageseq=1 (pp. 489–490)

Facts about the four UK nations: en.wikipedia.org/wiki/Countries_of_the_ United_Kingdom

Demographics of the European Union: en.wikipedia.org/wiki/Demographics_ of_the_European_Union

WWF-UK and Wild Ingleborough: www.wwf.org.uk/wild-ingleborough

Are the wildlife NGOs doing a good enough job?

My friend Rob Sheldon seems to have coined the term 'peak NGO' – at least he did as far as I was concerned. And I spent some time telling him that we hadn't got to that state yet – but time and experience have persuaded me that he was spot on.

Heal Rewilding: www.healrewilding.org.uk

Tarras Valley nature reserve (the Langholm Initiative): www.langholminitiative. org.uk/tarrasvalleynaturereserve

The UK Parliament website on petitions in history: erskinemay.parliament.uk/ section/5072/a-brief-history-of-petitioning-parliament

How to start a petition to the UK parliament (and about issues in England): petition.parliament.uk/petitions/check

How to start a petition to the Scottish parliament: www.parliament.scot/ get-involved/petitions/about-petitions

How to start a petition to the Welsh Senedd: petitions.senedd.wales/help

A petition calling for an end to Badger culls which received 108,320 signatures: petition.parliament.uk/archived/petitions/165672

A petition calling for the banning of driven grouse shooting which received 123,077 signatures: petition.parliament.uk/archived/petitions/125003

The extreme poverty of the NGO world

The 28 and their incomes in 2020/21 (all figures in millions):

England and Wales: register-of-charities.charitycommission.gov.uk/charity-search

Scotland: www.oscr.org.uk/search

A Rocha UK £1.0, Amphibian and Reptile Conservation £2.2, Badger Trust £0.3, Bat Conservation Trust £1.6, Buglife £1.6, Butterfly Conservation £4.5, Campaign for National Parks £0.3, Freshwater Habitats Trust £1.6, John Muir Trust £4, Mammal Society £0.2, Marine Conservation Society £4.3, National Trust £508, National Trust for Scotland £44m, People's Trust for Endangered Species £1.3, Plantlife £4.1, Rewilding Britain £0.8, Rivers Trust £3.9, Royal Society for the Protection of Birds £142, Scottish Beaver Trust £<<0.1, Trees for Life £3, Wild Fish Conservation £0.7, Shark Trust £0.5, Whale and Dolphin Conservation £5.0, Wildfowl and Wetland Trust £21, Wild Justice £0.2, all Wildlife Trusts £205, Woodland Trust £61 (£87 over a 17-month changed accounting period), World Wide Fund for Nature £84.

The combined income of all 28 organisations was £1,029 million.

We are few

Wildlife and Countryside Link's membership organisations have a total of over 8 million members: www.wcl.org.uk/our-members.asp – with the main contributors as follows:

National Trust, 5.4 million: www.nationaltrust.org.uk/lists/fascinating-facts-and-figures

RSPB, 1+ million: www.rspb.org.uk/join-and-donate/join-us-today

Wildlife Trusts, 0.9 million: www.wildlifetrusts.org/about-us

Woodland Trust, 0.5 million: www.woodlandtrust.org.uk/about-us/our-history

Wildfowl and Wetlands Trust, 0.18 million: www.wwt.org.uk/about-us/annual-reports

RSPB legacy income: register-of-charities.charitycommission.gov.uk/charity-search/-/charity-details/207076/financial-history

UK Army has 112,000 soldiers: www.army.mod.uk

Premiership football matches attendances, 2021/2022 season: www.statista.com/statistics/268576/clubs-of-the-english-premier-league-by-average-attendance

Liverpool FC season ticket prices: www.liverpoolfc.com/tickets/seasonticketrenewals/prices

Membership of political parties: commonslibrary.parliament.uk/research-briefings/sn05125

Someone I have only met in person once, at a Hen Harrier Day rally in the Peak District, called Dennis Ames, commented frequently on blogs I wrote for the RSPB back in 2009–10 and then on my own blog between 2011 and 2021, and he often made the point that the RSPB might have a million members but that left 66 million non-members. It took a while for me to appreciate how right he was.

Chapter 6: What wildlife needs (and how to provide it)

Alternative visions for a better future
Open Seas: www.openseas.org.uk
Land sharing and sparing: Green, R.E., Cornell, S.J., Scharlemann, J.P.W. & Balmford, A. 2005. Farming and the fate of wild nature. *Science* 307: 550–555. www.science.org/doi/abs/10.1126/science.1106049

Proposal 1 – Let's have an effective protected-area network
IUCN Crossroads blog, 20 August 2021. We need to protect and conserve 30% of the planet: but it has to be the right 30%: www.iucn.org/crossroads-blog/202108/we-need-protect-and-conserve-30-planet-it-has-be-right-30
UK Prime Minister, press release, 20 September 2020. PM commits to protect 30% of UK land in boost for biodiversity: www.gov.uk/government/news/pm-commits-to-protect-30-of-uk-land-in-boost-for-biodiversity
Starnes, T., Beresford, A.E., Buchanan, G.M. *et al.* 2021. The extent and effectiveness of protected areas in the UK. *Global Ecology and Conservation* 30: e01745. doi.org/10.1016/j.gecco.2021.e01745
European sites in Scotland: www.nature.scot/professional-advice/protected-areas-and-species/protected-areas/international-designations/european-sites
The National Sites Network in England and Wales: naturalresources.wales/about-us/news-and-events/blog/natura-2000-day-life-wrb
British Ecological Society. 2022. *Protected Areas and Nature Recovery: Achieving the Goal to Protect 30% of UK Land and Seas for Nature by 2030.* BES, London. www.britishecologicalsociety.org//wp-content/uploads/2022/04/BES_Protected_Areas_Report.pdf
National Nature Reserves in England: www.gov.uk/government/collections/national-nature-reserves-in-england
Where to look up on a map the various types of designated wildlife sites (and a bunch of other fascinating information) near you: magic.defra.gov.uk/MagicMap.aspx
Where to look up the site condition and last assessment date for your favourite English SSSI: designatedsites.naturalengland.org.uk
The Scottish equivalent compendium of MAGIC, giving locations, designations and whether the features are in good nick or not: www.nature.scot/professional-advice/protected-areas-and-species/protected-areas/national-designations/sites-special-scientific-interest-sssis

Proposal 2 – Increase public land ownership

A list of SSSIs in Northamptonshire: en.wikipedia.org/wiki/List_of_Sites_of_Special_Scientific_Interest_in_Northamptonshire

Wightman, A. 2013. *The Poor Had No Lawyers: Who Owns Scotland (and How They Got It)*. Birlinn Books, Edinburgh. And see www.andywightman.com/archives/category/who-owns-scotland

Cahill, K. 2001. *Who Owns Britain: the Hidden Facts Behind Landownership in the UK and Ireland*. Canongate Books, Edinburgh.

Shrubsole, G. 2019. *Who Owns England? How We Lost Our Green and Pleasant Land and How to Get It Back*. Bloomsbury, London; reviewed by me here: markavery.info/2019/05/05/sunday-book-review-who-owns-england-by-guy-shrubsole (and see also www.whoownsengland.org)

A brief history of the Forestry Commission: www.woodlands.co.uk/blog/woodland-economics/a-brief-history-of-the-forestry-commission

Examples of government compulsory purchase powers: www.gov.uk/guidance/compulsory-purchase-and-compensation-guide-4-compensation-to-residential-owners-and-occupiers

Water supply (and with it, sewerage services) has gone in and out of public ownership over the years. Water was provided, and sewage removed, as a public health provision by the state from the start of the twentieth century onwards without metering and supplied according to need rather than means first by local authorities and then, from the 1970s, by regional water authorities which were privatised by the Thatcher government in 1987. With that privatisation a large area of publicly owned and managed land was lost to the state too. The Labour election manifesto of 2017 proposed to renationalise water supply, and if that had come to pass then large areas of upland England would now be in state control and could be managed for wildlife, carbon, flood alleviation and recreation, as well as for water quality. Labour has said it no longer intends to take water utilities back into public ownership (which, personally, I think is a shame): www.mirror.co.uk/news/politics/keir-starmer-sticks-plan-nationalise-27566893

Proposal 3 – Rewild the uplands

A good introduction to rewilding projects in Scotland from Mossy Earth: mossy.earth/rewilding-knowledge/rewilding-scotland

Wild Alladale: alladale.com/rewilding

Glen Affric. Trees for Life: treesforlife.org.uk/about-us/affric-highlands – and in the words of Rewilding Britain: www.rewildingbritain.org.uk/local-network/affric-highlands

Mar Lodge. National Trust for Scotland: www.nts.org.uk/visit/places/mar-lodge-estate – and in the words of Rewilding Britain: www.rewildingbritain.org.uk/rewilding-projects/mar-lodge

Cairngorms Connect: cairngormsconnect.org.uk

Tarras Valley nature reserve (the Langholm Initiative): www.langholminitiative.org.uk/tarrasvalleynaturereserve

Monbiot, G. 2013. *Feral*. Allen Lane, London; reviewed by me here: markavery. info/2013/07/14/book-review-feral-george-monbiot

Tree, I. 2018. *Wilding*. Thames & Hudson, London; reviewed by me here: markavery.info/2018/07/29/sunday-book-review-wilding-by-isabella-tree

Macdonald, B. 2019. *Rebirding*. Pelagic Publishing, Exeter; reviewed by me here: markavery.info/2019/03/03/sunday-book-review-rebirding-by-benedict-macdonald

Painting, A. 2021. *Regeneration*. Birlinn, Edinburgh; reviewed by me here: markavery.info/2021/02/28/sunday-book-review-regeneration-by-andrew-painting

Schofield, L. 2022. *Wild Fell*. Doubleday, London; reviewed by me here: markavery.info/2022/02/20/book-review-wild-fell-by-lee-schofield

Macdonald, B. 2022. *Cornerstones*. Bloomsbury, London; reviewed by me here: markavery.info/2022/07/03/sunday-book-review-cornerstones-by-benedict-macdonald

Proposal 4 – Get tough with unsustainable farming

This is a well-worn subject and we are always just about to move to a better relationship with farming, but it never seems to happen. After the foot-and-mouth outbreak of 2001 Don Curry was asked to produce a report on the Future of Farming and Food, which he and his team did: webarchive. nationalarchives.gov.uk/ukgwa/20091112174647/http://archive.cabinetoffice. gov.uk/farming/pdf/PC%20Report2.pdf. That report is still worth a read and its concerns are still largely relevant.

Area of farmed land in the UK in 2019 according to a large land agency firm: www.savills.co.uk/research_articles/229130/274017-0

Area of organic farming in UK in 2021 according to UK government: www.gov. uk/government/statistics/organic-farming-statistics-2021

Trends in UK land area farmed organically over the last two decades: www.statista. com/statistics/298986/organic-land-used-in-the-united-kingdom-uk

Wheat as an animal food: www.veterinariadigital.com/en/articulos/importance-of-wheat-in-animal-feed-and-production

Where I get my weekly organic vegetable box – from Riverford: www.riverford. co.uk/ethics-and-ethos

Proposal 5 – Get political to change the system

My current MP is a Conservative, Tom Pursglove, a man whose political beliefs I do not share, but I often write to him just to let him know what I care about. To find how to contact your political representative in parliament see:

www.parliament.uk/get-involved/contact-an-mp-or-lord/contact-your-mp for Westminster MPs

www.parliament.scot/msps for Members of the Scottish Parliament

senedd.wales/find-a-member-of-the-senedd for Members of the Senedd

www.niassembly.gov.uk/your-mlas for Northern Ireland Assembly Members

Size of UK electorate, 2021: www.ons.gov.uk/peoplepopulationandcommunity/ elections/electoralregistration/bulletins/electoralstatisticsforuk/december2021

Proposal 6 – Choose the best NGOs

Environmental Funders network. 2022. What the Green Groups Said 2021: www.
greenfunders.org/what-the-green-groups-said-2021. Rankings of *The 28* by
votes cast: RSPB (22), Wildlife Trusts (20), National Trust (9), Woodland
Trust (8), Plantlife (7), Buglife (6), Rewilding Britain (5), WWF-UK (5),
Rivers Trust (4), Wild Justice (4), Butterfly Conservation (3), Marine Conservation Society (3), Bat Conservation Trust (2), Wildfowl and Wetland
Trust (2), Amphibian and Reptile Conservation (1), National Trust for
Scotland (1), Peoples Trust for Endangered Species (1) and all others (0).
Note that three organisations that in Chapter 3 I didn't class as conserva-
tion organisations score lowly here: BTO 1 vote (but if it were a ranking for
science then it would score very highly), BASC 0 votes, GWCT 1 vote.

Buglife: www.buglife.org.uk

Butterfly Conservation: www.butterfly-conservation.org

Open Seas: www.openseas.org.uk

Plantlife: www.plantlife.org.uk

Rewilding Britain: www.rewildingbritain.org.uk

Trees for Life: www.treesforlife.org.uk

Wild Justice: www.wildjustice.org.uk

Proposal 7 – Be an active investor

Take a look at the annual reports and accounts of some wildlife conservation
organisations and judge for yourself whether they give you a good picture
of how they are doing. You'll find that they provide strong confirmation
of the views expressed at the start of Chapter 2 that one can judge what
really matters in life by what is counted carefully. Wildlife conservation organisations are required to produce vast amounts of detailed summary of
where the money came from and how it was invested and spent, but little
information on whether they achieved anything. Embedded in the financial
information you will find the salaries of the organisation's top staff, and you
will find them interesting, and seeing them you may feel more emboldened
to be an active investor and ask some questions.

RSPB Annual Report 2021–2022: www.rspb.org.uk/globalassets/downloads/
annual-report-2022/annual-report-2021-22_digital_full_final.pdf. Very slick
and very glossy, but a bit thin on real information about conservation
progress (when compared with the other two organisations below). The
two pages (pp. 8–9) entitled 'What we do' seem particularly fuzzy rather
than demonstrating any clarity and are very badly written. Considering the
RSPB manages a land area larger than a smallish county, the analysis of
the fate of wildlife on that land area is cursory, mainly consisting of a few
'good news' stories. I think the RSPB seriously undersells its conservation
work in this document and looks more like a club than the campaigning
organisation that is needed, although the report does promise a strategic
shift to 'Being a bolder and more influential campaigning organisation' (p.
10). Which is very welcome (if delivered). Worryingly, this report makes

little mention of the role of land purchase in the RSPB's view of future conservation action. Later, there are 15 pages on 'governance', which are quite well written but unbearably dull. Overall, the RSPB's annual report makes it look like a big business, with much of the accompanying jargon, which does some nature conservation. I know that isn't what the RSPB is really like. Its annual report should do a much better job.

Wildlife Trust of Bedfordshire, Cambridgeshire and Northamptonshire Annual Report 2020–2021: www.wildlifebcn.org/about-us/how-were-funded/ annual-reports-and-accounts. Considering this is an organisation with an annual income of £7 million, its annual report is well judged and quite informative. It brings home the message that my money is spent on local projects, but I knew that anyway, although there really is very little reference to what the Wildlife Trust family is doing together – which might make you wonder about how successful the Wildlife Trusts are on the national stage. This Wildlife Trust is notably stronger on monitoring and research than most others, even the bigger ones, and that comes through in this report. Because of reporting requirements well over half of the 65 pages are about finances.

Woodland Trust Annual Report and Accounts January 2020 – May 2021: www. woodlandtrust.org.uk/media/50671/woodland-trust-report-accounts-jan20-may21.pdf. I'd never looked at the Woodland Trust report before as I am not a member, but maybe I should be. This is an organisation that is definitely about trees and woodlands – that's what they talk about here and they report quite well on outcomes, although there is a lack of information on how species have fared on the large area of land that they manage. Well over half of the 127 pages comprise the statutory reporting, mostly about money. The Woodland Trust has changed its accounting period, so this report covers 17 months and £87 million income (which pro rata is c. £61 million per annum). I was quite impressed by this report except for the fact that it talks in detailed terms about woodland coverage but only in vague terms about whether that woodland cover is actually delivering much wildlife.

I'd be really interested to hear whether you take the step of reviewing your investment in wildlife NGOs, as described in Proposals 6 and 7. Please let me know if you do, and tell me how you get on: mark@markavery.info. What responses did you get? If you are willing then I'll publish summaries on my website so that others can be informed.

Acknowledgements

This book is a product of my thoughts and so I should thank everybody who has helped to shape them over the years and I do, whether they be parents, children, friends, colleagues, teachers, lecturers, people whose talks I've heard, people on the radio or television, people who have commented on my blog in arresting ways, people whose books I've read and everyone else. But these are my reflections and I take full responsibility for them.

My wife, Rosemary, is very patient and very understanding.

Nigel Massen, David Hawkins and Sarah Stott of Pelagic Publishing and my editor and friend, Hugh Brazier, have been very patient too. This book has been more of a struggle to write than any other book I've written, perhaps because I have been writing about the issues that have occupied most of my life and about which I care deeply. If you come across a phrase in these pages and think 'That's quite good' then sometimes it will be because I got it right first time, and sometimes it will be because Hugh got me to get it right eventually. Of course, if you never have that thought, then that's completely my responsibility.

Index

Dr Mark Avery is a senior UK conservationist with nearly four decades' experience of giving wildlife a better future. The author of numerous previous books, including *Inglorious: Conflict in the Uplands* (2015), Mark worked for the RSPB for 25 years before going freelance in 2011. He co-founded the campaigning organisation Wild Justice (with Chris Packham and Ruth Tingay) and was recently chair of the World Land Trust. He lives in rural Northamptonshire where he tries to grow tomatoes and to add bird species to his garden list - both with limited success.

Twitter: @markavery

Also available from Pelagic

Birds & Flowers: An Intimate 50 Million Year Relationship (coming spring 2024), Jeff Ollerton

Traffication: How Cars Destroy Nature and What We Can Do About It, Paul F. Donald

Reconnection: Fixing Our Broken Relationship with Nature, Miles Richardson

Treated Like Animals: Improving the Lives of the Creatures We Own, Eat and Use, Alick Simmons

Invisible Friends: How Microbes Shape Our Lives and the World Around Us, Jake M. Robinson

Low-Carbon Birding, edited by Javier Caletrío

The Hen Harrier's Year, Ian Carter and Dan Powell

Wildlife Photography Fieldcraft: How to Find and Photograph UK Wildlife, Susan Young

Rhythms of Nature: Wildlife and Wild Places Between the Moors, Ian Carter

Ancient Woods, Trees and Forests: Ecology, Conservation and Management, edited by Alper H. Çolak, Simay Kırca and Ian D. Rotherham

Essex Rock: Geology Beneath the Landscape, Ian Mercer and Ros Mercer

Wild Mull: A Natural History of the Island and its People, Stephen Littlewood and Martin Jones

Challenges in Estuarine and Coastal Science, edited by John Humphreys and Sally Little

A Natural History of Insects in 100 Limericks, Richard A. Jones and Calvin Ure-Jones

pelagicpublishing.com

CPSIA information can be obtained
at www.ICGtesting.com
Printed in the USA
BVHW040017160623
665973BV00002B/5